B.K.A.

ESSENTIALS OF BIOSTATISTICS

ESSENTIALS OF BIOSTATISTICS

2nd Edition

ROBERT C. ELSTON, Ph.D.
Case Western Reserve University
Department of Epidemiology and Biostatistics
MetroHealth Medical Center
Cleveland, Ohio

WILLIAM D. JOHNSON, M.S.
Associate Professor
Department of Biometry and Genetics
Louisiana State University Medical Center
New Orleans, Louisiana

F.A. DAVIS COMPANY • Philadelphia

F. A. Davis Company
1915 Arch Street
Philadelphia, PA 19103

Printed in the United States of America

Last digit indicates print number: 10 9 8 7 6 5 4 3 2

Acquisitions Editor: Robert W. Reinhardt
Developmental Editor: Bernice M. Wissler
Production Editor: Arofan Gregory
Cover Design By: Steven Ross Morrone

As new scientific information becomes available through basic and clinical research, recommended treatments and drug therapies undergo changes. The authors and publisher have done everything possible to make this book accurate, up to date, and in accord with accepted standards at the time of publication. The authors, editors, and publisher are not responsible for errors or omissions or for consequences from application of the book, and make no warranty, expressed or implied, in regard to the contents of the book. Any practice described in this book should be applied by the reader in accordance with professional standards of care used in regard to the unique circumstances that may apply in each situation. The reader is advised always to check product information (package inserts) for changes and new information regarding dose and contraindications before administering any drug. Caution is especially urged when using new or infrequently ordered drugs.

Library of Congress Cataloging-in-Publication Data

Elston, Robert C., 1932–
Essentials of biostatistics / Robert C. Elston, William D. Johnson.—Ed. 2.
p. cm.
Includes bibliographical references and index.
ISBN-0-8036-3123-5
1. Biometry. 2. Medical statistics. I. Johnson, William Davis, 1941– . II. Title.
QH323.5.E47 1993
574′.01′5195—dc20
93-31126
CIP

PREFACE TO THE SECOND EDITION

The enthusiasm with which the first edition of our book was received by medical and nursing students has encouraged us in preparing this second edition. An additional pleasant response to the first edition came from many students of statistics and colleagues who commented on the book's usefulness as a survey of statistical methods. The basic philosophy behind the book remains the same, as explained in the Epilogue (page 275): our aim is to explain the basic and fundamental concepts without getting bogged down in computational details.

The major changes and additions since the first edition are in Chapters 2, 4, 7 and 12. Chapter 2 contains more material on the design of clinical trials. In Chapter 4 we explain Bayes' theorem pictorially; we thank Dr. Max P. Baur of Bonn University for introducing this pedagogic method to us. In Chapter 7 we make a clear distinction between significance tests and tests of hypotheses, and briefly introduce the notion of a posterior probability that a hypothesis is true. Chapter 12 contains a new section on meta-analysis, which is seen more and more in the medical literature. In addition, we have moved the discussion of standardized variables from Chapter 6 to Chapter 5. We have added a table of the normal distribution, which many instructors were unhappy to find missing in the first edition. We have increased the number of problems at the end of the chapters from 10 to 15, so that the book now contains 200 tested problems. In our experience, students find answering such problems an excellent way to learn. All the problems conform to the new format used in the National Board examinations.

Throughout the book we have made corrections, used student suggestions to clarify difficult concepts, and included mention of several further statistical tests that occur in the medical literature. Because we have not wanted to enlarge the book substantially, in many cases the additional information is limited to giving only the names and purposes of these tests.

This book has been successfully used as the required text both for a concentrated 16 lecture-hour course for medical students and for a 3 semester-hour introductory course for other health professionals. In the former case we have omitted Chapter 10 (Analysis of Variance and Linear Models) and covered the rest of the book in sufficient detail to enable the students to do the problems. The pace has necessarily been fast, but tolerably so. About 1 lecture hour is spent on each of Chapters 1, 2, 3, 5, 11 and 12, and about 2 lecture hours on each of Chapters 4, 6, 7, 8 and 9. The students are advised to read the appropriate material both before and after the corresponding lecture, and to use the chapter summaries as notes.

Finally, we wish to thank the many students and colleagues whose thoughtful comments and corrections have contributed to this second edition. We have taken great pains to eliminate errors that crept into the first edition unnoticed, but bear full responsibility for any errors that remain.

Robert C. Elston, M.A., Ph.D.
William D. Johnson, M.S.

PREFACE TO THE FIRST EDITION

"Biostatistics, far from being an unrelated mathematical science, is a discipline essential to modern medicine—a pillar in its edifice"[Journal of the American Medical Association (1966) 195:1145]. Today, even more so than twenty years ago, anyone who wishes to read the biomedical literature intelligently needs to understand the essentials of biostatistics. It is our hope that this book will provide such an understanding to those who have little or no statistical background and who need to keep abreast of new medical findings.

Unlike many other elementary books on biostatistics, the main focus of this book is not so much on teaching how to perform some of the simpler statistical procedures that may be necessary for a research paper, but rather on explaining basic concepts. Many of the simpler statistical procedures are in fact described, but computational details are included in the main body of the text only if they help clarify the underlying principles. For those who want them, more detailed computational formulae are given in Appendix 1. Similarly, we have relegated to Appendix 1 other details that, if included in the body of the text, would tend to make it difficult for the reader to see the forest for the trees. If you wish to have the details, read the notes in Appendix 1 concurrently, chapter by chapter, with the rest of the book.

This book has been written at an elementary mathematical level and requires no more than high school mathematics to understand. Nevertheless, you may find Chapters 4 and 5 a little difficult at first. These chapters on probability and distributions are basic building blocks for subsequent concepts, however, and you should study them carefully. The basic concepts of estimation and hypothesis testing are covered by the end of Chapter 7, and the next four chapters cover special statistical methods that are widely used in biological and medical research. In the last chapter we have tried to review the most important concepts introduced in earlier chapters as they relate to a critical reading of reports published in the biomedical literature.

We have attempted to illustrate the statistical methods described with enough examples to clarify the principles involved, but without their being so many and so detailed that the reader is caught up in irrelevant and unnecessary technicalities. We have tried to make these examples realistic and yet easy to grasp for someone with a background in the biological sciences. Many of our examples come from human genetics. Apart from providing ideal examples in the application of probability, genetics is a discipline that underlies all biology, and so should be of interest to the readers of our book. Detailed knowledge of genetics is not, however, necessary to understand the examples.

Many persons have helped in the preparation of this book and it would be impossible to mention them all by name. Special thanks, however, go to Dr. Robert M. Rippey, Dr. Martin Kotler, Dr. James H. Godbold, Dr. Beth Dawson-Saunders, and numerous medical students at LSU who read a draft of the manuscript and offered many suggestions for correction and improvement, and to Ms. Gina M. Drake and Ms. Elizabeth A. Gallmann who painstakingly typed the manuscript. Any errors that remain, however, are solely our responsibility.

Robert C. Elston, M.A., Ph.D.
William D. Johnson, M.S.

CONTENTS

ESSENTIALS OF BIOSTATISTICS

CHAPTER ONE

Key Concepts

biostatistics
deductive and inductive reasoning
scientific method
statistic, statistics

Introduction: The Role and Relevance of Statistics in Medicine

WHAT IS BIOSTATISTICS?

When taking a clinical history, conducting a physical examination, or requesting laboratory analyses, radiographic evaluations, or other tests, a physician is collecting information (data) to help choose diagnostic and therapeutic actions. The decisions reached are based on knowledge obtained during training, from the literature, from experience, or from some similar source. General principles are applied to the specific situation at hand in order to reach the best decision possible for a particular patient. This type of reasoning—from the general to the specific—is called **deductive reasoning.** Much of basic medical training centers around deductive reasoning.

If it has not happened already, at some point in your training you must ask yourself: How do we obtain the information about what happens in general? We are told, for example, that patients with hypertension eventually have strokes if their blood pressure is not controlled, but how did we obtain this information in the first place? Does the rule always hold? Are there exceptions? How long can the patient go untreated without having a stroke? Just how high can the blood pressure level be before the patient is in imminent danger?

These questions are answered by "experience." But how do we pyramid the knowledge we glean from experience so that we do not make the same mistakes over and over again? We save the information gathered from experience and refer to it to make better judgments as we are faced by the need to make new decisions. Moreover, we conduct experiments and comparative studies to focus on questions that arise in our work. We study a few patients (or experimental animals), and from what we observe we try to make rational inferences about what happens in general. This type of reasoning—from the specific subject(s) at hand to the general—is called **inductive reasoning.** This approach to medical research—pushing back the bounds of knowledge concerning human health—follows what is known as the **scientific method,** which has four basic steps:

1. Making observations—i.e., gathering data.
2. Generating a hypothesis—the underlying law and order suggested by the data.
3. Deciding how to test the hypothesis—what critical data are required?
4. Experimenting (or observing)—this leads to an inference that either rejects or affirms the hypothesis. If the hypothesis is rejected, then we go back to step 2. If it is affirmed, this does not necessarily mean it is true, only that in light of current knowledge and methods it appears to be so. The hypothesis is constantly refined and tested as more knowledge becomes available.

It would be easy to reach conclusions on the basis of observations, were it not for the variability inherent in virtually all biological data. One of the most common decisions a health professional must make is whether an observation on a patient should be considered normal or abnormal. Is a particular observation more typical of a person with disease or of a person without disease? Is the observation outside the range typically found in a healthy person? If the patient were examined tomorrow, would one obtain essentially the same observation? Obviously, observations such as blood pressure evaluations vary greatly, both at different times on the same patient and from patient to patient. Clinical decisions must be made with this variability in mind.

Inductive inference is a much riskier procedure than deductive inference. In mathematics, we start with a set of axioms. Assuming that these axioms are true, we use **deductive** reasoning to prove things with certainty. In the scientific method, we use **inductive** inference and can never prove anything with absolute certainty. In trying to generalize results based on a group of 20 patients, you might ask such questions as: If 20 additional patients were studied, would the results be very close to those obtained on studying the first 20 patients? If a different laboratory analyzed the blood samples, would the results be similar? If the blood samples had been stored at a different temperature, would the results be the same?

A statistic (plural: statistics) is an estimate of an unknown numerical quantity, such as the mean height of men age 20. **Statistics** (singular) is a science that deals with the collection, organization, analysis, interpretation, and presentation of information that can be stated numerically. **Biostatistics** is statistics applied to the biological sciences. Perhaps the most difficult aspect of statistics is the logic associated with inductive inferences, yet all scientific evidence is based on this type of statistical inference. The same logic is used, though not always explicitly, when a physician practices medicine: what is observed for a particular patient is integrated with what has previously been observed for a large group of patients to make a specific decision about that particular patient.

REASONS FOR STUDYING BIOSTATISTICS

Many students ask: Why should I study biostatistics? or: Why would a health professional need to know statistics? The statement "If I need a statistician, I will hire one," is also common. But health professionals are frequently faced with data on which they must base clinical judgments. The reliability of support data plays a fundamental role in making good clinical decisions. You must be able to distinguish between discrepant data and routine variability. As a layperson and as a practitioner, you will be bombarded daily with statistics. To make correct decisions based on the data you have, you must know where those data came from and how they were obtained; you must also know whether conclusions based on those data are statistically valid. Statistics are often misinterpreted. Disraeli is reputed to have said, "There are lies, damned lies, and statistics."

You must be able to understand and evaluate the medical literature in an intelligent manner. Unfortunately, many of the articles in the medical literature draw invalid conclusions because incorrect statistical arguments are used. Schor and Karten (1966) found that most analytical studies published in well-respected medical journals in 1964 were unacceptable in that the conclusions drawn were not valid in terms of the design of the experiment, the type of analysis performed, or the applicability of the statistical tests used. Unfortunately, things were only slightly better 15 years later: Glantz (1980) reported that about half of the articles published in medical journals that use statistics use them incorrectly. After you complete your training, you will be relying on the literature for new information that will change the way you practice your profession. It is important that you be able to read these articles critically. You will need to understand terms such as "p-value," "significance level," "confidence interval," "standard deviation," and "correlation coefficient," to mention just a few of the statistical terms that are now common in the medical literature. This book explains these concepts and gives you an explanation of statistics that will help you distinguish fact from fancy in everyday life—in newspapers and on television, and in making daily comparisons and evaluations.

The easy availability of computers means that increasingly sophisticated data analysis will be performed to advance medical science. The computer augments cerebral functions in the same way that other machines augment muscular functions: the computer can receive, store, select, evaluate, and transmit information. To understand the meaning of computer output, you will need to understand the principles underlying the computations performed by the computer. Remember, however, that no amount of computation can produce information that is not already inherently present in a set of data. Computer scientists refer to this principle as GIGO—Garbage In, Garbage Out!

Finally, you should have an appreciation of statistics so that you know when, and for what purpose, a statistician should be consulted in medical research.

HOW CAN A STATISTICIAN HELP AN INVESTIGATOR?

Statistics is a vital component of the research process, from the earliest planning stages of a study to the final presentation of its results. The involvement of a statistician in all stages of a research project enhances the efficiency of the study and the scientific credibility of its results. If a study has been improperly planned or executed, no amount of statistical expertise can salvage its results. At the beginning of a study, the statistician's activities might include (1) recommending study designs to meet the objectives and to increase the amount of information that can be obtained; (2) helping develop efficient data-collection forms that are easily processed; and (3) recommending ways to monitor the quality of the data as they are being collected. After the data have been collected and prepared for analysis, the statistician can (1) recommend the most appropriate methods of analysis, and possibly do the analyses personally if the methods are sophisticated; (2) interpret the findings in understandable terms; and finally (3) review and contribute to the statistical content of any presentations and publications that report the results of the study. Statisticians may also be consulted to help evaluate published papers and manuscripts, or to help prepare sections on experimental design and analysis for grant applications.

WHAT DOES THE STATISTICIAN EXPECT FROM AN INVESTIGATOR?

Because statistical consultation is a professional collaboration, statisticians should be included in research projects from their beginning. Also, because statistical design and analysis are time-consuming activities, statisticians should be informed well in advance of any deadlines. At the beginning of the study, you should: (1) show the statistician exactly where and how the data will be collected, preferably at the collaboration site; (2) describe the objectives of your study in detail, because they are essential in planning a study design that will extract all pertinent information as efficiently as possible; and (3) inform the statistician of any relevant limitations, such as availability of financing or personnel, which are all factors that must be considered in the study design. To clarify the terminology and basic concepts in your field of interest, you should provide the statistician with background information in the form of basic articles and book chapters in your area of research.

Once a strategy for your study has been jointly established, it must be followed carefully, and any necessary changes in procedures must be discussed before they are implemented. Even minor changes may affect the way data are analyzed, and in some cases could invalidate an entire study. Although certain statistical methods may partially correct mistakes in study design, this is not always possible; it is obviously expedient to avoid making such mistakes at all.

In return for substantial contributions to a study, the statistician could reasonably expect to be a coauthor on publications reporting the results of the study, and should, of course, have the opportunity to review each publication before it is submitted. In the case of a minor contribution, an acknowledgment may be sufficient, but remember that it is as discourteous to acknowledge someone's help without that person's permission as it is to fail to make an appropriate acknowledgment. The question of how the statistician's contribution will be recognized should be brought up early in a project.

SOME SPECIFIC EXAMPLES

We are all aware of the accuracy of projections made by the media in predicting the outcome of national elections before many people have even gone to the polls to cast their vote. This process is based on a relatively small but representative sample of the population of precincts throughout the country. It is a classic example of statistical inference—drawing conclusions about the whole population based on a representative sample of that population.

Several years ago the King Tut art exhibit was on display in New Orleans. During the last few days of the exhibition, people waited in line for hours just to enter to see it. On the very last day, the lines were exceptionally long and seemed to be moving terribly slowly. One ingenious man decided to estimate his expected waiting time as follows. He stepped off 360 paces (approximately 360 yards) from his position to the front of the line. He then observed that the line moved 10 paces in 15 minutes. He projected this to estimate a movement of 40 paces per hour or 360 paces in 9 hours. The man then decided that 9 hours was too long a period to wait in line. A man (one of the authors of this book) directly behind this fellow, however, decided to wait and stood in line 9½ hours before seeing the exhibit!

In 1948, a report appeared of what turned out to be a classic study of streptomycin and tuberculosis. Patients were assigned to a streptomycin group or a control group and assessed after 6 months. Fourteen of 52 patients in the control group died during the 6-month period, compared with 4 of 55 patients in the streptomycin group. Moreover, among those who lived, significantly more patients in the streptomycin group showed considerable improvement on radiological evaluation. It was inferred from this, and subsequent experience has provided even stronger evidence, that streptomycin is an effective treatment for tuberculosis.

Today most health experts believe that smoking is bad for the health—that it increases the risk of cancer and heart attacks, and has other deleterious effects on the body. Governments have taken actions to modify or suppress advertising by the tobacco industry, and to educate the public about the harmful effects of smoking. These actions were taken, however, only after many independent studies collected statistical data and drew similar conclusions.

SUMMARY

1. Biostatistics deals with the collection, organization, presentation, analysis, and interpretation of biological information that can be stated numerically.
2. All data collected from biological systems have variability. The statistician is concerned with summarizing trends in data and drawing conclusions in spite of the uncertainty created by variability in the data.
3. Deductive reasoning is reasoning from the general to the specific. Inductive reasoning is drawing general conclusions based on specific observations.
4. The scientific method provides an objective way of formulating new ideas, checking these ideas with real data, and pyramiding findings to push back the bounds of knowledge. The steps are as follows: make observations, formulate a hypothesis and a plan to test it, experiment, and then either retain or reject the hypothesis.
5. An understanding of statistics will enhance your ability to interpret data, whether for the purpose of treating a particular patient or for drawing general conclusions from a research study, as well as enable you to distinguish fact from fancy in everyday life.

FURTHER READING

Everitt, B.S. (1989) Statistical Methods for Medical Investigations. Oxford University Press. (Chapter 1 of this book gives a good introduction to some of the important considerations in conducting medical studies.)

Glantz, S.A. (1980) Biostatistics: How to Detect, Correct and Prevent Errors in the Medical Literature. Circulation 61:1–7. (You will gain more from this article if you read it after you have learned biostatistics.)

Huff, D. (1954) How to Lie with Statistics. Norton, New York. (Everyone who has not read this book should do so: you can read it in bed!)

Medical Research Council. (1948) Streptomycin Treatment of Pulmonary Tuberculosis. British Medical Journal 2:769–782.

Schor, S.S., and Karten, I. (1966) Statistical Evaluation of Medical Journal Manuscripts. Journal of the American Medical Association 195:1123–1128.

CHAPTER TWO

Key Concepts

cause and effect

adverse drug reaction

confounding

target population, study population, study unit, census, parameter

probability sample:
random cluster sample
simple random sample
stratified random sample
systematic random sample
two-stage cluster sample

observational study:
cohort/prospective study
case-control/retrospective study
historical cohort/historical prospective study
matched pairs
sampling designs

experimental study:
completely randomized
fractional factorial arrangement
randomized blocks
split-plot design
changeover/crossover design
sequential design

factorial arrangement of treatments

response variables, concomitant variables

longitudinal studies, growth curves, repeated measures studies, follow-up studies

clinical trial, placebo effect, blinding, masking

double blinding, double masking

compliance, adherence

Populations, Samples, and Study Design

THE STUDY OF CAUSE AND EFFECT

Very early in your study of science you probably learned that if you put one green plant in an area exposed to sunlight and another green plant in a dark area, such as a closet, the plant in the dark area would turn yellow after a few days, whereas the one exposed to sunlight would remain green. This observation involves a simple experiment in which there are just two plants. Can we infer from it that we have a **cause** (light) and an **effect** (green color)? Will the same thing happen again? We might put several plants in sunlight, and several in a closet, and we might repeat the experiment on many different occasions, each time obtaining the same result. This would convince us that what we observed was not pure coincidence, but can we be sure that it is the light that causes the green color (or, conversely, darkness that causes the yellow color)? If the closet is always cooler than the sunlit areas, the color change could be simply due to the cooler temperature. Clearly such an experiment—in which temperature is not carefully controlled—cannot distinguish between lack of sunlight and cooler temperature as the cause of the color change. In this situation, we say there is **confounding**—the effect of light is confounded with that of temperature. Two factors in a study are said to be confounded when it is not possible to determine from the study which factor is causing the effect being investigated. In this chapter, we are going to discuss ways of designing studies with appropriate safeguards against confounding and other pitfalls, so that we can be more certain about making correct inferences about causes and effects.

Much of medicine is concerned with the study of disease—identifying the cause of disease, preventing disease by intervention, and treating disease to minimize its impact. Unfortunately, most diseases have a complex pathogenesis, and so it is not easy to describe the underlying process. In the simplest situation, factor A causes disease X. In more complex situations, disease X is caused by multiple factors, say factors A, B, C, and D, or factor A may cause more than one disease. Factor A may cause disease X only in the presence of factor B, so that if either factor A or factor B is present alone, no causative effect can be observed. In another instance, factor A could initiate disease X, but factor B could accelerate or promote the disease process once it is initiated. On the other hand, disease X could influence factor A, whereas factor A might have no influence on the disease. The time of exposure to possible causative factors is another consideration. Some patients might have more than one disease at the same time, and these diseases might be associated with some of the same causative factors. Moreover, it is not always obvious that a disease is present, especially in the preclinical or early clinical stages.

Because of these many possibilities, the determination of disease causation may be complex, and therefore it is essential that studies be carefully designed. Above all, it is important to realize that, in our attempt to push back the bounds of knowledge, we are searching for truth. Truth remains constant and it has a way of making itself evident. The investigator must therefore be objective; he must not discard data that are inconsistent with his expectations or otherwise manipulate findings under the guise of saving the time required for additional investigation. All too often we are misled by reported findings only to discover later that a researcher was not careful and objective in his research.

In any good scientific study the objectives will be clearly stated, including specific hypotheses to be tested and effects to be estimated. For example, the objectives might be to

1. Identify a group of men with elevated serum cholesterol levels.
2. Reduce the serum cholesterol levels in these men by administering a treatment, say drug A.
3. Determine whether a reduction in serum cholesterol levels over a 5-year period reduces the risk of developing coronary heart disease.
4. Determine whether drug A has adverse side effects that outweigh any reduction in the risk of coronary heart disease.

Thus the specific hypothesis to be tested is that using drug A over a 5-year period, to lower the serum cholesterol levels of men with elevated levels, reduces their risk of developing coronary heart disease. The specific effects to be estimated are the amount of reduced risk and the amount of adverse reaction to the drugs. We note that a side effect is not necessarily adverse. Moreover, adverse experiences may be caused by factors other than the experimental treatment. **Adverse drug reactions** include such symptoms as nausea, vomiting, diarrhea, abdominal pain, rash, drowsiness, insomnia, weakness, headache, dizziness, muscle twitching, and fever.

At the outset, definitions of terms should be specified. For example, coronary heart disease might be defined as myocardial infarction or a history of angina, but this definition in turn requires definitions of myocardial infarction and angina. We also need to define what is meant by an "elevated" serum cholesterol level. These considerations will determine which men enter the study, and hence the men for whom results of the study will have relevance. If the definitions of these terms are not clearly documented, it might be difficult to apply consistent criteria in deciding which patients are eligible for study. If patients who are free of coronary heart disease are entered into the study, the effects of a treatment aimed at this disease will be diluted. Similarly, if only patients with the most severe disease are entered, the effect of treatment as it applies to the general population of patients might be exaggerated. We now turn to a general discussion of how the inferences we wish to make are limited by the manner in which our study units are selected. (In the example we have just considered, each "study unit" is a man with an elevated serum cholesterol level.)

POPULATIONS, TARGET POPULATIONS, AND STUDY UNITS

One of the primary aims of statistics is to help draw objective conclusions that pertain to a larger group than the one for which data are available. You might want, for example, to compare two treatments for angina. Obviously, you cannot give the treatments to all patients with angina, but you can study a small group of such patients. You would then try to generalize your conclusions based on the results from that small group to a larger group of patients—perhaps even to most patients with angina. Similarly, it would not be practical to study all men with elevated serum cholesterol levels, but we would hope to learn something relevant to all such men on the basis of an experiment performed on a small group, or sample.

In these examples, the set of all angina patients, or the set of all men with elevated cholesterol levels (the word "elevated" being precisely defined), would be the **population** about which we wish to make inferences. The population is made up of **study units,** i.e., the units we study from the population—which in these examples are angina patients or men with elevated cholesterol levels, respectively. Although it seems obvious in these examples that we are studying people, and that these people make up specific populations, the situation is not always this clear. The population of interest may, for example, be made up of blood samples or tissue specimens, each such sample or specimen being a study unit. We often use animal models in our initial investigations of an area of human research, and so the study unit might be a rat, a hamster, or a dog.

The important point to remember is that in statistics, a population is the group of all study units about which a particular investigation may provide information. The study units make up the population, and the population about which we wish to make inferences determines what is meant by a study unit. Suppose, for example, that we are interested in the functioning of the muscle cells of patients with myotonic dystrophy. We might take two muscle biopsies from each of five such patients and divide each biopsy into three aliquots, or parts, making a total of 30 aliquots in all, on each of which we make a measurement. But, if the population about which we wish to make inferences is the population of patients with myotonic dystrophy, our sample contains only five study units—on each of which we have the measurements made on 6 aliquots. If we wish to make inferences about all the (theoretically) possible biopsies that could be taken from one particular patient, then each biopsy is a study unit and we have a sample of only two such units on any one patient. It is a common error to believe that increasing the number of measures taken on any one study unit is equivalent to increasing the number of study units sampled from a population. If our experiment with the green plants had involved only two plants—one in the light and one in the dark—then, since we were interested in making inferences about plants, only two study units were in the experiment. This number is not increased to 20 by noting that each plant has 10 leaves and then recording the color of each leaf. We must carefully distinguish between multiple study units and multiple measures on a single study unit—a distinction that is intimately tied to the population about which we wish to make inferences.

We must also carefully distinguish between the **target population** and the **study population.** The target population is the whole group of study units to which we are interested in applying our conclusions. The study population, on the other hand, is the group of study units to which we can legitimately apply our conclusions. Unfortunately the target population is not always readily accessible, and we can study only that part of it that is available. If, for example, we are conducting a telephone interview to study all adults (our target population) in a particular city, we do not have access to those persons who do not have a telephone. We may wish to study in a particular community the effect of drug A on all men with cholesterol levels above a specified value; however, short of sampling all men in the community, only those men who for some reason visit a doctor's office, clinic, or hospital are available for a blood sample to be taken. Thus, we have a study population of accessible study units and a target population that includes both the study population and the inaccessible study units. Those study units that are not readily accessible may or may not have the same characteristics as those of the study population. Their exclusion from the study population means that inferences made about the study population need not necessarily apply to the target population.

There are many ways to collect information about the study population. One way is to conduct a complete **census** of the population by collecting data for every study unit in it. The amount of money, time, and effort required to conduct a complete census is usually unreasonable. A more practical approach is to study some fraction, or sample, of the population. If the sample is representative of the population, then inferences we make from the sample data about the population will be correct. The term **statistic** is used to designate a quantity computed from sample data, and the term **parameter** is used to designate a quantity that is characteristic of the population. If the sample is representative of the population, descriptive statistics will give accurate impressions of the corresponding parameters of the population. Since our interest is in estimating parameters and testing hypotheses about parameters of the population, special efforts should be taken to obtain a representative sample. Haphazard samples, or samples selected on the basis of being easy to collect, are rarely representative of the population. We will now describe methods of sampling a population that usually lead to representative samples, and that, in any case, allow us to make valid inferences.

PROBABILITY SAMPLES AND RANDOMIZATION

In order to make the kinds of inferences we discuss in later chapters, we select well-defined **probability samples,** in which every unit in the population has a known probability of being included in the sample. The most elementary type of probability sample is the **simple random sample,** in which every study unit in the population is equally likely to be included in the sample. However, it is often better to take another kind of probability sample. Suppose, for example, we take a simple random sample of all individuals in a community, and it just happens, by chance, that there are no women in our sample. Gender might be an important factor in what we are studying, and if this is the case, we would have obtained a sample that may be seriously deficient. To overcome this possibility, we take a **stratified random sample**—a sample obtained by separating the population study units into nonoverlapping groups, called strata, and then selecting a simple random sample from each stratum. Thus, the population is first divided by gender (the two strata), and a random sample then taken from each stratum. In this way we can ensure that each gender is represented in the sample in proportion to its distribution in the population. Similarly, we might stratify the population on the basis of age, socioeconomic status, health status, and so forth, before taking a simple random sample of a predetermined size from each stratum.

One approach to selecting a random sample is to put the name, number, or some other identification of each study unit in the population into a container, such as a hat, mix the identifications thoroughly, and then select units (i.e., identifications) one at a time until the required sample size has been attained. This procedure, however, has practical limitations. It may not be possible to obtain a thorough mixing, and it may be impractical to write down the identification of every unit in the population on a separate piece of paper. However, provided we have a list of these identifications (such as a list of all hospital patients who, during a period of time have a particular disease, or a list of all residents in a community), a sequence of random numbers can be used to pick the sample. Thus, if our list contains 10,000 names, a 1% random sample can be selected by obtaining a sequence of 100 random numbers between 1 and 10,000, and using those numbers to indicate the positions in the list of those persons who are to form the sample. Tables of random numbers have been published for this purpose. We would start at an arbitrary place in the table and then take as our random numbers the next 100 sets of four digits that appear successively in the table. The four digits 0000 would be interpreted as 10,000, and in the unlikely event that the same number is obtained twice, another number would be taken. There are also computer programs that generate pseudo-random numbers that appear to be random, but are in fact produced by a well-defined numerical algorithm. The fact that they do not produce truly random numbers is usually of little consequence when a single sample is being selected for a particular study.

A simpler way to select a sample from a list is to take what is called a **systematic random sample.** Here we randomly select a number between 1 and k—where 1/k is the fraction of study units in the population we wish to have in our sample—and then select every k^{th} unit in the list. This type of design is often used in selecting patients from a large clinic on the basis of hospital charts. We might study every 1 in 20 charts, for example, simply selecting a random number between 1 and 20, say 16, and studying the 16th, 36th, 56th, and so forth, chart. The disadvantage of this kind of sample is that the order of the study units on the list may have a periodic pattern, in which situation we may obtain an unrepresentative sample. Provided we can be sure that this is not the situation, however, a systematic random sample is a conveniently simple way to obtain an approximately random sample.

Often the study units appear in groups, and so we take a random sample of clusters or a **random cluster sample.** The physicians working in a hospital could be considered a cluster. Thus, we could sample hospitals (clusters) and interview every physician working in each hospital selected in the sample. A **two-stage cluster sample** is obtained by first selecting a sample of clusters (stage one) and then selecting a sample of study units within each cluster (stage two). In this situation, we really have two different study units, since the clusters themselves can be considered as individual study units in the population of all such clusters. As we shall see in Chapter 10, there are special methods of analysis for this kind of situation.

OBSERVATIONAL STUDIES

So far, we have considered the problem of sampling study units from the population, and we have seen that different ways of doing this, or different **sampling designs,** are possible. The sampling design is usually critical to the interpretation of **observational studies,** that is, studies in which the researcher merely "observes" the study units, making one or more measurements on each.

There are many types of observational studies, e.g., to determine the prevalence of disease, or to determine the population distribution of a given trait, such as blood pressure. Two types of observational studies that we shall discuss in more detail provide different epidemiological approaches to investigate the cause of disease. The first approach is to identify a sample of persons in whom the suspected cause is present and a second sample in whom the cause is absent, and then compare the frequency of development of disease in the two samples. Studies that proceed in this manner are called **cohort,** or **prospective, studies.** A cohort is a group of people who have a common characteristic; the two samples are thus cohorts, and we observe them "prospectively" for the occurrence of disease. We might, for example, identify one cohort of smokers and one cohort of nonsmokers, and then observe them for a period to determine the rate of development of lung cancer in the two groups.

The other approach is to identify a sample of patients who have the disease of interest (**cases**) and a second sample of persons who do not have the disease (**controls**), and then compare the frequency of possible causes of the disease in the two groups. These are called **case-control,** or **retrospective, studies.** They are retrospective in that we "look back" for possible causes. Thus we might identify a group of lung-cancer patients and a group of lung-cancer–free subjects, and then study their past history of smoking.

Thus, the difference between cohort and case-control studies is in the selection of the persons (the study units) for study. In a cohort study, persons are selected who are initially free of disease, and we determine disease frequency over some period of time, in the presence or absence of factors suspected of being associated with causing the disease. On the other hand, in a case-control study, persons are selected on the basis of presence or absence of the disease, and we determine the frequency of possible causative factors in their past histories. In both types of observational studies there is always the possibility that confounding factors are present. It is known, for example, that smokers tend to drink more coffee than nonsmokers. Thus, any association found between smoking and lung cancer could in theory merely reflect a causal link between coffee drinking and lung cancer. Provided the potential confounding factors can be identified, methods of analysis exist to investigate this possibility. The major drawback to observational studies is that there may be unidentified confounding factors.

In cohort studies, persons are usually selected at the time the study is started and then followed over time. Existing records alone, however, can be used to conduct a cohort study. The groups are established on the basis of possible causative factors documented at an early point in the records, and disease frequency is then established over a subsequent period in the same existing records (e.g., work records at an industrial plant). Such a study, even though it is conducted retrospectively, is called a **historical cohort,** or **historical prospective, study.** The important point is that the outcome (disease or not) is determined prospectively in relation to the time in the record at which the samples are chosen.

The source of cases in most case-control studies is provided by the patients with the disease of interest who are admitted to a single hospital or to a group of hospitals during a specific interval of time. The controls are often patients admitted to the same hospitals during the same interval of time for reasons other than and unrelated to the disease under study. Alternatively, instead of such hospitalized controls, an effort may be made to obtain population-based controls. The type of controls used can play a critical role in the interpretation of the results, since hospitalization per se (regardless of a person's status with respect to the disease being studied) may be a confounding factor related to the putative causative factors.

To obtain a group of controls that is comparable to the cases, the controls are often matched to the cases with respect to extraneous variables. Thus, for example, as a particular case is entered into the study, one or more persons of the same sex and race, and of similar age and socioeconomic status, are identified and entered into the control group. We say that the cases and controls have been matched for sex, race, age (within a specified interval, e.g., within 5 years) and socioeconomic status. A **matched pairs** design, implying that there is one control matched to each case, is common.

The case-control study design is frequently used to explore simultaneously a number of possible causes of disease. Such a study can usually be conducted from existing records and is therefore relatively inexpensive. Furthermore, it can be conducted in a reasonably short time. Often the purpose of a case-control study is to learn enough about the causes of a disease of interest to narrow the range of possibilities. Then a prospective study, with a small number of clearly defined hypotheses, can be planned and conducted.

EXPERIMENTAL STUDIES

In **experimental studies**, the researcher intervenes ("experiments") in some way to affect the manner in which the study units (in this case often called "experimental units") respond. The study units are given some stimulus, or treatment, and then a response is observed. For example, our study units could be patients with the common cold, and the treatment could be a drug that we believe may be a cure for the common cold. We give the drug to a sample of such patients, and 2 weeks later observe that virtually none of them has a cold any more. This is, in fact, an experimental study, but does it allow us to conclude that the drug is a cure for the common cold? It clearly does not, because we would expect the same result even if the patients had not received the drug. We all know that common colds tend to go away after a couple of weeks, whether or not any drugs are taken. This example illustrates the very simple and obvious principle that if a result follows a particular course of action, it does not mean that the course of action necessarily causes the result. Studies of treatment efficacy are often plagued by ignoring this simple principle—that in many cases, time alone is sufficient to "cause" a cure or an amelioration of a disease.

For time to be eliminated as a possible confounding factor, experimental studies must have at least two treatment groups. The groups are compared with each other after the same amount of time has elapsed. If there are two groups, one will be given the treatment being investigated, and the other, the control group, may be given no treatment. We say there are two different treatment groups, since "no treatment" is in itself a type of treatment. Thus, we say there are two possible treatments: drug therapy versus no drug therapy. Or, if it would be unethical not to treat the patient at all, the two treatments could be a new drug we are investigating and a control treatment that has been (until now) the standard treatment for such patients. There may be more than two groups—one being given the new treatment and two or more others being given two or more other standard treatments for the disease. The way in which the study units are assigned to the various treatment groups, together with any relationships there may be among the different treatments, determines what is known as the **experimental design.**

Careful **randomization** is an essential component of any sound experimental design. Ideally, all of the study units in an experiment should be obtained from the target population using an element of randomization. But this is usually impractical, and the sample of study units used is often not at all representative of the target population. Nevertheless, randomization plays an important role in the allocation of study units to the various treatments of experimental studies. Suppose, for example, that patients with a specific disease are assigned to one of four treatments, and that after a suitable period of observation, the treatment groups are compared with respect to the response observed. Although the patients studied might be representative of those seen at a particular clinic, they are unlikely to be representative of all patients with the disease of interest (the target population). But if the patients are assigned to the treatments in such a way that each patient is equally likely to receive any one of them, a fair comparison of the four treatments is possible. Thus, the patients should be randomly assigned to the treatment groups, as this will enhance the chances of a fair distribution of the good- and poor-risk patients among the four groups. If 32 patients are available, they can each be given a number, the 32 numbers written on separate pieces of paper, shuffled, and then randomly sorted into four treatment piles. A table of random numbers could also be used for the same purpose.

It cannot be stressed too much that without a proper randomization procedure, biases—wittingly or unwittingly—are almost bound to result. If we had 32 mice in a cage to be assigned to four different treatments, it would not be sufficient simply to reach in and pick out eight "at random" for the first treatment, eight "at random" for the second, and so on. An obvious bias occurs with this procedure in that the last group will consist of those mice that are less easy to catch, and the first group of those easiest to catch. Random allocation of study units to the treatment groups is an automatic safeguard against possible confounding factors. This is an essential difference between any observational study and a well-conducted experimental study (i.e., an experimental study in which there is random allocation of study units to treatments).

Many types of experimental designs are available, all involving an appropriate element of randomization. The main reason for choosing one design over another is to save the experimenter money, time, and effort, while at the same time ensuring that the objectives of the study can be met. It will be helpful to review very briefly a few of the most common experimental designs. **A completely randomized design** is one in which each study unit has the same probability of being assigned to any treatment group under consideration. Thus, if three treatments, A, B, and C, are being investigated, then each study unit would have a 1/3 probability of being assigned to any one of the three treatment groups.

A **randomized blocks design** is one in which the study units are first divided into nonoverlapping "blocks" (sets of units, or strata) and then randomly assigned to the treatment groups separately within each block. Consider, for example, an experiment in which a group of patients is separated by gender to form two blocks. The male patients are randomly assigned to the treatment groups (using a completely randomized design approach), and then the female patients are separately randomly assigned to the treatment groups. In this way, if patients of one gender always tend to give a better response, this "sex effect" can be "blocked out" in the sense that treatments are compared within the blocks of male and female patients. Provided the treatment *differences* are similar among the male and among the female patients, it is possible to pool the results from the two sexes to obtain an overall comparison of treatments that is not obscured by any sex effect.

The **split-plot design** involves randomly assigning units ("whole plots") to treatment groups and then randomly assigning subunits ("split plots") to a second kind of treatment group. (The use of the term "plot" arose from agricultural experiments in which each experimental unit was a plot in a field.) For example, patients (whole plots) with advanced diabetes might be randomly assigned to different treatments for diabetes and then their eyes (the split plots) might be randomly assigned to different treatments to correct eyesight. As a second example, female mice (whole plots) might be randomly exposed to some experimental drug and their offspring (split plots) might be randomly assigned to a second treatment.

The **changeover, or crossover, design** is used, especially in animal and human studies, in an effort to use the study unit (patient) as his or her own control. In this design, units are randomly assigned to one treatment for a period, and then switched to another treatment for a second period. For example, if two treatments, say A and B, are being investigated, one group of patients would be assigned to treatment A for, say 2 weeks, and then to treatment B for 2 weeks. A second group of patients would be assigned to treatment B for 2 weeks followed by treatment A for 2 weeks. In this design, a residual or carryover effect of the treatment given in the first period may affect the results found in the second period. In an attempt to prevent this, a rest (or "washout") period is sometimes given between the two treatments. Provided there is no carryover effect, or if the carryover effect is the same when B is followed by A as when A is followed by B, there is no problem in using all the data to estimate the true difference between the effects of the two treatments. If this is not the case, difficulties arise in analyzing the results of this kind of experimental design.

In many clinical studies, patients are recruited into the study over weeks, months, or even years. Whenever there is a lapse in time between the study of successive experimental units, a **sequential design** may be used. In such a design, the data are analyzed periodically as they are collected, and the result of each analysis determines whether to continue recruiting patients into the study or whether the study should be terminated and a decision made, on the basis of the data gathered so far, as to which treatment is best. It is also possible to use the result of each analysis to determine the probability of assigning the next patient to each of the treatments. In this way, as evidence accumulates that a particular treatment is best, the next patient recruited into the study has a greater probability of receiving that treatment. This strategy is called "playing the winner."

So far, we have discussed that aspect of experimental design that concerns the way in which the experimental units are assigned to different treatments. We have stressed that randomization must be involved at this step if we are to avoid biases and be able to make valid inferences. Another aspect of experimental design concerns the choice of the different treatments to investigate. We may be interested in the effects of three different doses of drug A—call these treatments A_1, A_2, and A_3. We may also be interested in the effects of two different doses of drug B, say B_1 and B_2, as treatments for the same disease. Furthermore, we may be interested in investigating whether there is any advantage in using these drugs in combination. We could set up separate experiments to investigate each of these three questions. But it is more economical to investigate the three questions simultaneously in a single experiment that has a **factorial arrangement** of the treatments. By this we mean that there are two or more factors of interest (two in our example—drug A and drug B), each at two or more levels (three levels of drug A and two levels of drug B in our example), and that our treatments comprise all possible combinations of different levels of each factor. Thus, in our example, there would be $3 \times 2 = 6$ different treatments, which we can label A_1B_1, A_1B_2, A_2B_1, A_2B_2, A_3B_1 and A_3B_2. Some investigators might be inclined to run three experiments for this example: one to study A_1 versus A_2 versus A_3, a second to study B_1 versus B_2, and a third to study combinations of the two drugs. This highly inefficient approach should be avoided. In some special circumstances, it may be feasible to study combinations of treatments without using all possible treatment combinations in the experiment. Such experiments are said to have a **fractional factorial** arrangement of the treatments.

It is important to keep in mind that both the choice of treatments and the way in which the study units are assigned to them (i.e., the method of randomization) determine the experimental design. Thus we could have a completely randomized design or any of the other experimental designs mentioned above with a factorial arrangement of treatments. It is also important to keep in mind that the purpose of the design is to have an experimental plan that answers questions of interest in as efficient a manner as is practical. Advanced concepts may be required for this purpose, and one of the roles of the consulting statistician is to provide an efficient design that meets the needs of the investigator.

There are many other general aspects of experimental design, only a few of which are mentioned here. Laboratory techniques should be refined to minimize, or perhaps even eliminate entirely, sources of extraneous variability, such as observer biases, measurement errors, and instrument variability. Where possible, large sources of variability should be used as a basis for defining "blocks" in a randomized blocks experiment. Sometimes it may be possible to measure, but not control with any accuracy, factors that could be important sources of variability. In a biochemical experiment in which each study unit is a reaction mixture, for example, it may be possible to control the temperature of each mixture to within 2°C, but no more accurately than that. On the other hand, it might be possible to measure the actual temperature attained in each mixture with great accuracy. If such small temperature changes could be critical to the outcome measures of primary interest, then the temperature of each mixture should be recorded. Measures that are thought to be affected by the different experimental conditions are those of primary interest and are often referred to as **response variables.** Other measures that are not themselves of primary interest, but may have an important effect on the response variable(s), are called **concomitant variables.** Thus temperature is a concomitant variable in our example, and a statistical technique called the analysis of covariance can make allowances for it; this technique is briefly discussed in Chapter 10. The precision of experimental results can often be greatly improved if appropriate concomitant variables are measured at the time of a study.

Finally, an important consideration in any investigation is the number of study units to include. The more study units observed, the more reliable our conclusions will be; however, we should like to obtain reliable results with a minimum amount of money, time, and effort. As we shall see in Chapter 7, statistical methods are available for estimating the number of study units required once a study design has been chosen.

We will now turn to special considerations of experimental design when the study units are human subjects.

Clinical trials are experimental studies that involve people as study units. Clinical trials have come to play a major role in deciding the efficacy of new drugs and other treatments as they become available. Early investigations of new treatments tend to focus on animal studies, but the ultimate evaluation involves a clinical trial. Often the response of each study unit is observed on two or more occasions. These investigations are called **longitudinal studies.** In these investigations it is sometimes of interest to model changes with time in terms of mathematical functions or **growth curves.** A distinction is sometimes made between longitudinal studies, in which the response is observed over long periods, and **repeated measures studies,** in which data are collected over a relatively short period of time—frequently under experimental conditions that change over time, as in the changeover design. Another special type of longitudinal study is called a **follow-up study.** In follow-up studies, the response outcome is the time to occurrence of some endpoint such as death, disease, or remission of disease. Because of the difficulties in maintaining an experiment over long periods of time, there is a greater chance of having missing data in long-term follow-up studies. Incomplete or missing data add to the complexities of statistical analysis.

The process of developing a new drug usually begins in a research laboratory in preclinical studies. As the drug is developed, animal studies are conducted. The experimental drug is introduced to humans in *phase I trials,* which involve about 10 to 20 very closely monitored subjects. The purpose of this early testing is to regulate dose tolerance and drug action in people. Healthy adult male volunteers are often used as subjects in phase I trials, but patients with the disorder of interest are also used in some investigations. *Phase II trials* are conducted to determine the effectiveness and safety of the new drug, relative to another drug or a placebo, and to regulate further the preferred dose (may vary with disease severity). *Phase III trials* are conducted to demonstrate drug efficacy and safety in patients typical of those expected to use the drug. These trials usually involve a large number of subjects and several investigators, and the duration of the study is often lengthy. *Phase IV studies* are conducted to monitor long-term experience with the drug after it is marketed. Phase I, II, and III studies are conducted to support applications to the Federal Drug Administration for permission to market a new drug.

In the case of a clinical trial, randomization tends to provide a good distribution of both poor- and good-risk patients in all treatment groups. Obviously, if one treatment group were assigned only good-risk patients and the other only poor-risk patients, a subsequent comparison of treatment effects would be biased. But even if we achieve a perfect randomization to the different treatments, it is still possible to misinterpret the effective differences among the treatments. If one group of patients is given an injection of a drug and the other is not, we cannot tell whether the difference in outcome is caused by the drug itself or by the act of injection. (Recall the example of the green plant that turns yellow in the closet: without a proper control we cannot be sure whether this effect is caused by less light or less heat.) A better plan would be to inject the control patients in the same manner, with a similar fluid, but one that does not contain the active drug. In this way the act of injection is no longer a confounding variable.

To enhance objectivity in evaluating treatments in clinical trials, the patient and/or the evaluating clinician are often not told which treatment the patient is receiving. This is especially important if one of the treatments is the administration of an inactive treatment (such as an injection or a pill containing no active drug), called a **placebo.** It is not unusual to observe an improvement in a group of patients on placebo therapy (a **placebo effect**) when the patients do not know they are receiving a placebo; however, if they know they are receiving a placebo, the effect is destroyed. Withholding information about which treatment is being used is call **blinding** or **masking;** when both the patient and the evaluating clinicians are blinded (masked), the procedure is called **double blinding** or **double masking.**

Researchers involved in clinical trials often go to great lengths to try to enhance **compliance (adherence)** to a treatment regimen. If the treatment is a one-time treatment, compliance is not a problem; however, if it is a treatment in which the patient must take a prescribed drug one or more times a day, then compliance is often poor. Frequent contact (e.g., once a week) between the physician and the patient can enhance compliance but can also be prohibitively expensive. It may be helpful to provide motivational programs, including educational lectures and group discussions. Compliance can be monitored by counting the number of pills remaining in the container at each visit, or by using specially designed dispensers. Compliance may influence response and vice versa; therefore, bias may be introduced if the types of subjects not complying to study protocol or the reasons for noncompliance differ among treatment groups. Special analyses may be required to assess the nature of noncompliance and its impact on treatment comparisons.

An important consideration in clinical trials is ethics. If a physician believes that one treatment is better than another, can that physician ethically assign patients to treatments in a random fashion? On the other hand, is it ethical *not* to conduct a clinical trial if we are not sure which treatment is best? If the risks of adverse reactions or undersirable side effects are great, can the physician ethically prescribe a drug? If early in the trial it becomes evident that one treatment is preferable to another, can we continue to randomly assign patients to all the treatments? Questions such as these must be addressed by internal review boards in all clinical trials. Furthermore, the purpose of the study and the possible risks must be described in lay terms to the patient, who must then sign an "informed consent" form agreeing to participate in the study. But the patient must be given the option of leaving the study at any time, and if this option is exercised, care must be taken in analyzing the results of the trial to ensure that it introduces no serious biases.

SUMMARY

1. The investigation of cause and effect requires well-designed studies. That B follows A does not imply that A causes B, because confounding factors may be present.

2. Study units are sampled from a study population, which is usually only part of the target population of interest. The set of all possible study units makes up the study population. The study population about which we make inferences determines how the study units are defined. Multiple measurements made on a study unit do not increase the sample size.

3. Selection of samples from a population using an element of randomization allows one to draw valid inferences about the study population. If the population is heterogeneous, better representation is obtained by use of a stratified random sample. A systematic random sample is a convenient approximation to a random sample.

4. Two types of observational studies are used to investigate the causes of disease. In cohort, or prospective, studies, samples are chosen on the basis of the presence or absence of a suspected cause, and then followed over time to compare the frequency of disease development in each sample. A cohort, or prospective, study is termed historical if it is conducted totally on the basis of past records. In case-control, or retrospective, studies, samples are chosen on the basis of presence or absence of disease, and compared for possible causes in their past. The choice of controls in case-control studies is critical: they may be hospital- or population-based; matching for demographic and other factors is usually desirable.

5. In experimental studies, there is intervention on the part of the researcher, who subjects the study units to treatments. Randomization by an approved method in the allocation of study units to the different treatment groups provides a safeguard against possible confounding factors, so that valid inferences are possible.

6. In a completely randomized design, each study unit has equal probability of being assigned to any treatment. Heterogeneity among study units can be "blocked out" in a randomized blocks design. A split-plot design allows study units, after being assigned to treatments at a primary level, to be divided into subunits for assignment to treatments at a secondary level. In a changeover, or crossover, design, each study unit is subjected to two (or more) treatments over time and comparisons among treatments are made within study units. This design can lead to difficulties in the analysis and/or interpretation if there are carryover effects from one period to the next. In a sequential design, the data are analyzed periodically to determine whether and/or how to continue the study. A factorial arrangement of treatments is one in which the treatments comprise all possible combinations of different levels of two or more factors.

7. In all experiments, extraneous sources of variability should be kept to a minimum or blocked out. Any remaining variables that could have a large effect on the results but cannot be blocked out should be measured during the course of the experiment. Such concomitant variables can be used in an analysis of covariance to increase the precision of the results.

8. Clinical trials are experimental studies in which the study units are people. They are used to judge the efficacy of new drugs and other treatments. Great care is needed in the choice of the control or comparison (in view of the commonly found placebo effect) and in monitoring adherence to a regular regimen. Either the physician or the patient may be blinded, or masked, as to which treatment is being used. Ideally, both are blinded, in a "double-blinded" or "double-masked" trial. Ethical considerations play an important role in the design of clinical trials.

FURTHER READING

Altman, D.G. (1980) Statistics and Ethics in Medical Research: Study Design. British Medical Journal 281:1267–1269. (This is a succinct overview of the relation between design and ethics in observational studies and clinical trials.)

Marks, R.G. (1982) Designing a Research Project: The Basics of Biomedical Research Methodology. Wadsworth. (This is a book on the practical aspects of design for a novice researcher.)

Meinert, C.L. (1986) Clinical Trials: Design, Conduct and Analysis. Oxford University Press.

PROBLEMS

1. A physician decides to take a random sample of patient charts for the last 5 years at a large metropolitan hospital to study the frequency of cancer cases at that hospital. He estimates the number of charts to be 10,000 and decides to take a 5 percent sample (i.e., a sample of 500 charts). He decides to randomly select a number between 1 and 20 (e.g., suppose the number turned out to be 9) and then study every 20th chart beginning with that number (in this example, charts 9, 29, 49, 69, . . .). This is an example of a sample design known as a

   A. two-stage random sample
   B. stratified random sample
   C. systematic random sample
   D. random cluster sample
   E. simple random sample

2. A physician decides to take a random sample of hospitals in a large metropolitan area. From each hospital included in the sample he takes a random sample

of house-staff physicians. He interviews the physicians to determine where they attended medical school. The sample design used in this study is an example of a

A. systematic random sample
B. stratified random sample
C. simple cluster sample
D. two-stage random sample
E. haphazard sample

3. In a study of the cause of lung cancer, patients who had the disease were matched with controls by age, sex, place of residence, and social class. The frequency of cigarette smoking in the two groups was then compared. What type of study was this?

A. sample survey
B. experimental study
C. retrospective study
D. clinical trial
E. prospective study

4. Investigations in which the study units are stimulated in some way and the researcher observes a response are called

A. observational studies
B. prospective studies
C. sample surveys
D. experimental studies
E. retrospective studies

5. A study was undertaken to compare results of the surgical treatment of duodenal ulcer. A total of 1,358 patients who met the study criteria were randomly assigned to one of four surgical procedures. The purpose of the randomization was to

A. ensure that the double-blind aspect of the study was maintained
B. obtain the unbiased distribution of good- and poor-risk patients in all treatment groups
C. achieve the same number of patients on each operation
D. guarantee that the study group was a representative sample of the general population
E. remove the poor-risk patients from the study

6. Investigators who use nonrandomized controls in clinical trials often argue that their controls are satisfactory because the distribution of prognostic factors in the control and experimental groups is similar before therapy. It is better to randomize because

A. a placebo effect is easier to detect in randomized trials
B. randomization tends to magnify the differences between placebo and treatment groups
C. many important prognostic factors that were not considered may lead to bias
D. it is easier to maintain blinding in randomized trials
E. compliance is better in randomized trials

7. A researcher decides to conduct a clinical trial using 40 patients. She carries out her treatment assignment by a well-defined method of randomly allocating 20 patients to group I and 20 patients to group II. After 4 weeks on treatment A, patients in group I are taken off the treatment for 4 weeks, and then given treatment B for an additional 4 weeks. Similarly, patients in group II are given treatment B for 4 weeks, no treatment for four weeks, and then treatment A for 4 weeks. The design used in this study is called
A. a stratified design
B. a sequential design
C. a completely randomized design
D. a randomized block design
E. a changeover design

8. An experiment is to be conducted using a crossover design. The statistician informs the investigator that a rest period, or washout period, should be included in the study plan. The purpose of the rest period is to eliminate or reduce
A. observer bias
B. missing data
C. residual treatment effects
D. problems with patient compliance
E. adverse drug experiences

9. A study is conducted using either dose A_1 or A_2 of the drug A and dose B_1, B_2 or B_3 of drug B. The design was such that each patient was equally likely to receive any one of the treatment combinations A_1B_1, A_1B_2, A_1B_3, A_2B_1, A_2B_2 or A_2B_3. This is an example of a
A. randomized blocks design with a factorial arrangement of treatments
B. changeover design
C. completely randomized design with a factorial arrangement of treatments
D. sequential design
E. staggered design with a factorial arrangement of treatments

10. A combination drug has two components: A, the antihistamine, and B, the decongestant. A clinical trial is designed in two parts: part I to randomly assign

patients to either a placebo control or drug A, and part II to randomly assign a second group of patients to a placebo control or drug B. A more efficient plan is to use a design with a

A. two-stage cluster sampling
B. systematic random sample
C. good compliance history
D. random allocation of adverse drug experiences
E. factorial arrangement of treatments

11. Pregnant women were recruited into a drug trial during the 28th week of pregnancy. They were allocated at random and double-blinded to placebo or active treatment groups. This could be best defined as

A. a sample survey
B. a clinical trial
C. a retrospective study
D. a case-history study
E. an observational study

12. Studies involving patients randomly assigned to treatment groups and then observed to study response to treatment are called

A. retrospective studies
B. case-control studies
C. observational studies
D. clinical trials
E. sample surveys

13. In a double-blind, randomized trial of the effectiveness of a drug in the treatment of ulcers, patients were randomly assigned to either an active or a placebo group. Each person was followed up for 6 weeks and evaluated as showing: (1) significant improvement, or (2) no significant improvement. The purpose of the double-blind aspect was to

A. obtain a representative sample of the target population
B. achieve a good distribution of good- and poor-risk patients in the two groups
C. guard against observer bias
D. eliminate the necessity for a statistical test
E. ensure proper randomization

14. Studies using the changeover design are special types of

A. retrospective studies
B. repeated measures studies
C. observational studies

D. adverse drug reaction studies
E. intent-to-treat studies

15. In a randomized, double-blind clinical trial of an antihistamine drug versus a placebo control, the drug was found to be beneficial for the relief of congestion, but drowsiness, dizziness, jitteriness, and nausea were significantly more prevalent and more severe in the group receiving the drug. This is an example of
A. a changeover trial
B. adverse drug experiences which must be weighed against the merits of the drug
C. longitudinal data that require a growth curve interpretation
D. a trial in which compliance was not a problem
E. a study that should have been carried out using laboratory animals

CHAPTER THREE

Key Concepts

interval, ordinal, and nominal scale

quantitative, qualitative

continuous data, categorical or discrete data

table, frequency distribution

histogram, bar graph, frequency polygon, cumulative plot, scatter plot (scatter diagram), tree diagram, decision tree

proportion, percentage, rate

prevalence, incidence

relative risk, odds ratio, attributable risk

sensitivity, specificity, predictive values

measures of central tendency:
- *mean*
- *median*
- *mode*

measures of spread (variability):
- *range*
- *interquartile range*
- *variance*
- *standard deviation*
- *coefficient of variation*

skewness, kurtosis

Symbols and Abbreviations

AR	attributable risk
CV	coefficient of variation
g_2	fourth cumulant; the coefficient of kurtosis minus three (used to measure peakedness)
OR	odds ratio
RR	relative risk
s	sample standard deviation (estimate)
s^2	sample variance (estimate)

Descriptive Statistics

WHY DO WE NEED DESCRIPTIVE STATISTICS?

We stated in Chapter 1 that a statistic is an estimate of an unknown numerical quantity. A descriptive statistic is an estimate that summarizes a particular aspect of a set of observations. Descriptive statistics allow one to obtain a quick overview, or "feel," for a set of data without having to consider each observation, or datum, individually. (Note that the word "datum" is the singular form of the word "data"; strictly speaking, "data" is a plural noun, although, like "agenda," it is commonly used as a singular noun, especially in speech.)

In providing medical care for a specific patient, a physician must consider: (1) historical or background data, (2) diagnostic information, and (3) response to treatment. These data are kept in a patient chart that the physician reviews from time to time. In discussing the patient with colleagues, the physician summarizes the chart by describing the atypical data in it, which would ordinarily represent only a small fraction of the available data. To be able to distinguish the atypical data, the physician must know what is typical for the population at large. The descriptive statistics we discuss in this chapter are tools that are used to help describe a population. In addition, as we discussed in Chapter 2, researchers at universities and pharmaceutical companies conduct studies on samples of patients, and need to report their general findings to the medical community. Descriptive statistics are used to summarize the results of such studies in a succinct way. Tables and graphs are also useful in conveying a quick overview of a set of data, and in fact, tables and graphs are often used for displaying descriptive statistics. We therefore include a brief discussion of them in this chapter. First, however, we consider the different kinds of data that may need to be described.

SCALES OF MEASUREMENT

We are all familiar with using a ruler to measure length. The ruler is divided into intervals, such as centimeters, and this is called an **interval** scale. An interval scale is a scale that allows one to measure all possible fractional values within an interval. If we measure a person's height in inches, for example, we are not restricted to measures that are whole numbers of inches. The scale allows such measures as 70.75 or 74.5 inches. Other examples of interval scales are the Celsius scale for measuring temperature and any of the other metric scales. In each of these cases the trait that is being measured is **quantitative,** and we refer to a set of such measurements as **continuous** data. Height, weight, blood pressure, and serum cholesterol levels are all examples of quantitative traits that are commonly measured on interval scales. The number of children in a family is also a quantitative trait, since it is a numerical quantity; however, it is not measured on an interval scale, nor does a set of such numbers comprise continuous data. Only whole numbers are permissible, and such data are called **discrete.**

Sometimes we classify what we are measuring only into broad categories. For example, we might classify a patient as "tall," "medium," or "short," or as "hypertensive," "normotensive," or "hypotensive." The trait is then **qualitative,** and such measurements also give rise to **discrete,** or **categorical,** data consisting of the counts, or numbers of individuals, in each category. There are two types of categorical data, depending on whether or not there is a natural sequence in which we can order the categories. In the examples just given, there is a natural order: "medium" is between "tall" and "short," and "normotensive" is between "hypertensive" and "hypotensive." In this case the scale of measurement is called **ordinal.** The number of children in a family is also measured on an ordinal scale. If there is no such natural order, the scale is called **nominal,** with the categories having names only, and no sequence being implied. Hair color, for example (e.g., "brown," "blond," or "red"), would be observed on a nominal scale. Of course the distinction between a nominal and an ordinal scale may be decided subjectively in some situations. Some would argue that when we classify patients as "manic," "normal," or "depressed," this should be considered a nominal scale, while others say that it should be considered an ordinal one. The important thing to realize is that it is possible to consider categorical data from these two different viewpoints, with different implications for the kinds of conclusions we might draw from them.

TABLES

Data and descriptive statistics are often classified and summarized in **tables.** The exact form of a table will depend on the purpose for which it is designed as well as on the complexity of the material. There are no hard and fast rules for constructing tables, but it is best to follow a few simple guidelines to be consistent and to ensure that the table maintains its purpose:

1. The table should be relatively simple and easy to read.
2. The title, usually placed above the table, should be clear, concise, and to the point; it should indicate what is being tabulated.
3. The units of measure for the data should be given.
4. Each row and column, as appropriate, should be labeled concisely and clearly.
5. Totals should be shown, if appropriate.
6. Codes, abbreviations, and symbols should be explained in a footnote.
7. If the data are not original, their source should be given in a footnote.

Tables 3–1 and 3–2 are two very simple tables that display data we shall use, for illustrative purposes, later in this chapter. Table 3–1 is the simplest type of table possible. In it is given a set of "raw" data, the serum triglyceride values of 30 medical students. There is no special significance to the rows and columns, their only purpose being to line up the data in an

TABLE 3–1. **Fasting Serum Triglyceride Levels (mg/dl) of 30 Male Medical Students**

45	46	49	54	55
61	67	72	78	80
83	85	86	88	90
93	99	101	106	122
123	124	129	151	165
173	180	218	225	287

TABLE 3–2. **Frequency Distribution of Fasting Serum Cholesterol Levels (mg/dl) of 1,000 Male Medical Students**

Cholesterol Level (mg/dl)	Number of Students
90–100	2
100–110	8
110–120	14
120–130	21
130–140	22
140–150	28
150–160	95
160–170	102
170–180	121
180–190	166
190–200	119
200–210	96
210–220	93
220–230	35
230–240	30
240–250	23
250–260	15
260–270	7
270–280	3
280–290	1
Total	1,000

orderly and compact fashion. Note that, in addition, the values have been arranged in order from the smallest to the largest. In this respect the table is more helpful than if the values had simply been listed in the order in which they were determined in the laboratory. There is another kind of table in which the rows and columns have no special significance, and in which, furthermore, the entries are never ordered. This is a table of random numbers, which we referred to in Chapter 2.

The simplest type of descriptive statistic is a count, such as the number of persons with a particular attribute. Table 3–2 is a very simple example of how a set of counts can be displayed, as a **frequency distribution.** Each of the observed 1,000 cholesterol levels occurs in just one of the interval classes, even though it appears that some levels (e.g., 100 and 110 mg/dl) appear in two consecutive classes. Should a value be exactly at the borderline between two classes, it is included in the lower class. This is sometimes clarified by defining the intervals more carefully (e.g., 90.1–100.0, 100.1–110.0, etc.). Age classes are often defined as 0 to 9 years, 10 to 19 years, etc. It is then understood that the 10 to 19 year class, for example, contains all the children who have passed their 10th birthday but not their 20th birthday. Note that in Table 3–2 some of the information inherent in the original 1,000 cholesterol values has been lost, but for a simple quick overview of the data this kind of table is much more helpful than would be a table, similar to Table 3–1, that listed all 1,000 values.

GRAPHS

The relationships among numbers of various magnitudes can usually be seen more quickly and easily from graphs than from tables. There are many types of graphs, but the basic idea is to provide a sketch that quickly conveys general trends in the data to the reader. The following guidelines should be helpful in constructing graphs:

1. The simplest graph consistent with its purpose is the most effective. It should be both clear and accurate.
2. Every graph should be completely self-explanatory. It should be correctly and unambiguously labeled with title, data source if appropriate, scales, and explanatory keys or legends.
3. Whenever possible, the vertical scale should be selected so that the zero line appears on the graph.
4. The title is commonly placed below the graph.
5. The graph generally proceeds from left to right and from bottom to top. All labels and other writing should be placed accordingly.

One particular type of graph, a **histogram,** often provides a convenient way of depicting the shape of the distribution of data values. Two examples of histograms, relating to the data in Tables 3–1 and 3–2, are shown in Figures 3–1 and 3–2. The points you should note about histograms are

1. They are used for data measured on an interval scale.
2. The visual picture obtained depends on the width of the class interval used, which is to a large extent arbitrary. A width of 10 mg/dl was chosen for Figure 3–1, and a width of 20 mg/dl for Figure 3–2. It is usually best to choose a width that results in a total of 10 to 20 classes.

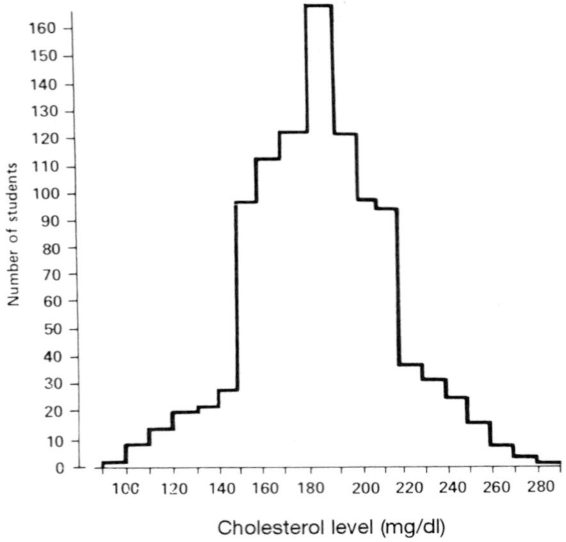

Figure 3–1. Histogram of 1000 fasting serum cholesterol levels (from Table 3–2).

3. If the observations within each class interval are too few, a histogram gives a poor representation of the distribution of counts in the population. Figure 3–2 suggests a distribution with several peaks, whereas a single peak would most likely have been found if 1,000 triglyceride values had been used to obtain the figure. More observations per class interval could have been obtained by choosing a wider interval, but

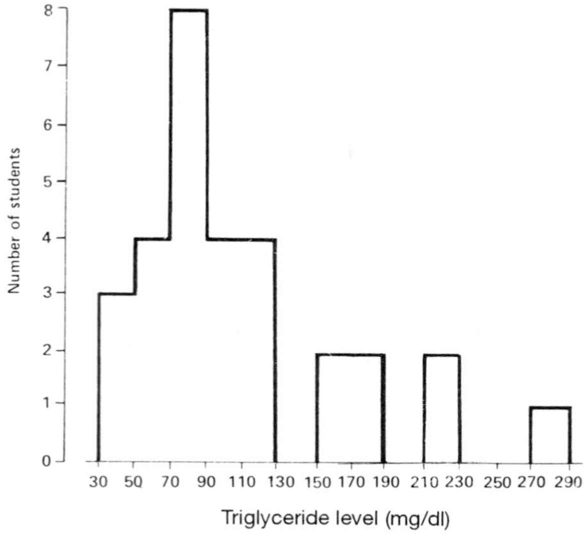

Figure 3–2. Histogram of 30 fasting serum triglyceride levels (from Table 3–1).

fewer than 10 intervals gives only a gross approximation to a distribution.

A **bar graph** is very similar to a histogram but is used for categorical data. It may illustrate, for example, the distribution of the number of cases of a disease in different countries. It would look very similar to Figures 3–1 and 3–2, but since the horizontal scale is not continuous, it would be more appropriate to leave gaps between the vertical rectangles or "bars." Sometimes the bars are drawn horizontally, with the vertical scale of the graph denoting the different categories. In each case, as also in the case of a histogram, the length of the bar represents either a frequency or a relative frequency, sometimes expressed as a percentage.

A **frequency polygon** is also basically similar to a histogram and is used for continuous data. It is obtained from a histogram by joining the midpoints of the top of each "bar." Drawn as frequency polygons, the two histograms in Figures 3–1 and 3–2 look like Figures 3–3 and 3–4. Notice that the polygon meets the horizontal axis whenever there is a zero frequency in an interval—in particular this occurs at the two ends of the distribution. Again the vertical scale may be actual frequency or relative frequency, the latter being obtained by dividing each frequency by the total number of observations; we have chosen to use relative frequency. A frequency polygon is an attempt to obtain a better approximation, from a sample of data, to the smooth curve that would be obtained from a large population. It has the further advantage over a histogram of permitting two or more fre-

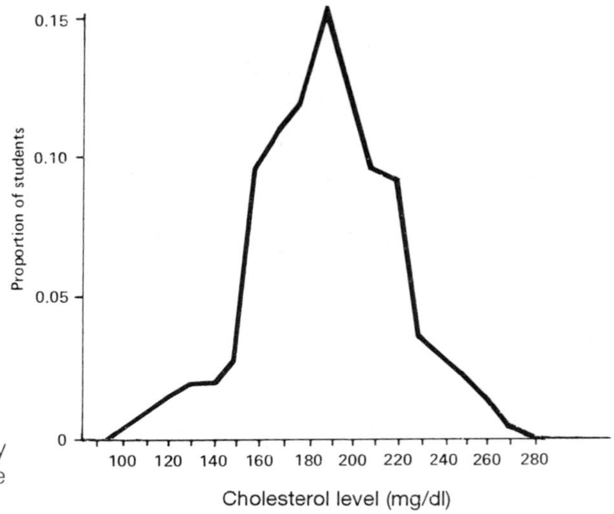

Figure 3–3. Relative frequency polygon corresponding to Figure 3–1.

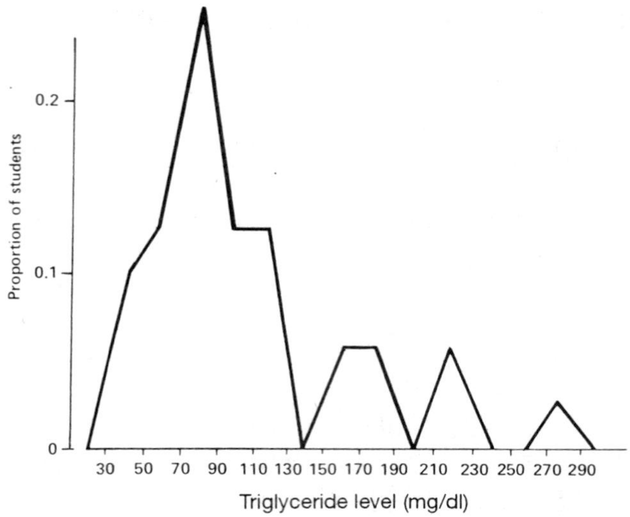

Figure 3–4. Relative frequency polygon corresponding to Figure 3–2.

quency polygons to be superimposed in the same figure with a minimum of crossing lines.

A **cumulative plot** is an alternative way of depicting a set of quantitative data. The horizontal scale (abscissa) is the same as before, but the vertical scale (ordinate) now indicates the proportion of the observations less than or equal to a particular value. A cumulative plot of the data in Table 3–2 is presented in Figure 3–5. We see in Table 3–2, for example, that $2 + 8 + 14 + 21 + 22 + 28 + 95 = 190$ out of the 1,000 students have serum cholesterol levels less than or equal to 160 mg/dl, and so the height of the point above 160 in Figure 3–5 is 190/1,000, or 0.19. We could similarly draw a cumulative plot corresponding to the histogram of the 30 triglyceride values (Figure 3–2), but one of the great advantages of the cumulative plot is that it does not require one to group the data into interval classes, as does a histogram. In a cumulative plot every single observation can be depicted, as illustrated in Figure 3–6 for the data in Table 3–1. It is clear from that table that 1 out of 30 values is less than or equal to 45, 2 out of 30 less than or equal to 46, 3 out of 30 are less than or equal to 49, and so forth. So we can make 1/30 the ordinate at 45, 2/30 the ordinate at 46, 3/30 the ordinate at 49, and so forth, up to $30/30 = 1$ as the ordinate at 287. However, the purpose of the cumulative plot is to approximate the continuous curve we would obtain with a much larger set of numbers. If more observations were included, one of them might possibly be larger than any of the values in Table 3–1. For this reason it is customary to make the ordinate at the largest data point (287 in this instance) some-

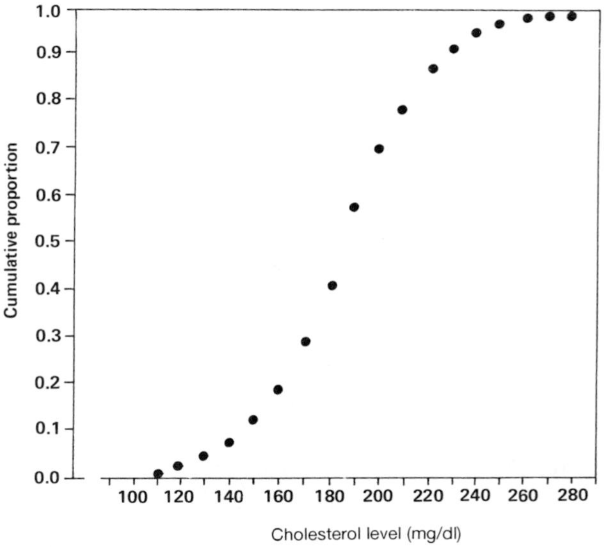

Figure 3–5. Cumulative plot of the data in Table 3–2.

what less than unity. One convenient way of doing this is to use one more than the total number of observations as the divisor. Thus the ordinates for the data in Table 3–1 are depicted in Figure 3–6 as 1/31 at 45, 2/31 at 46, 3/31 at 49, up to 30/31 at 287. Note that a cumulative plot results in a much smoother curve than the histogram (Fig. 3–2) and that all the information in the original table is retained.

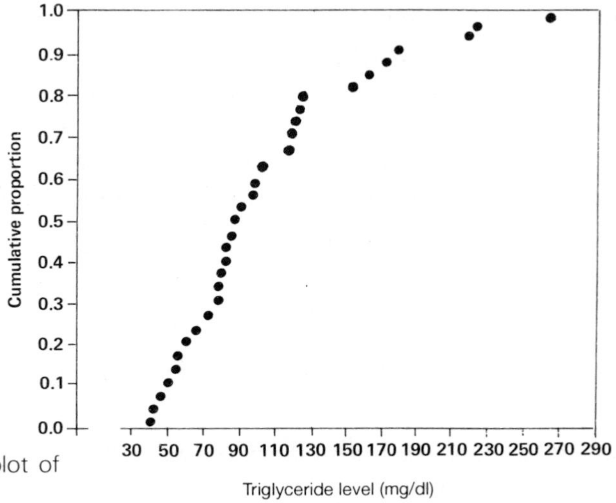

Figure 3–6. Cumulative plot of the data in Table 3–1.

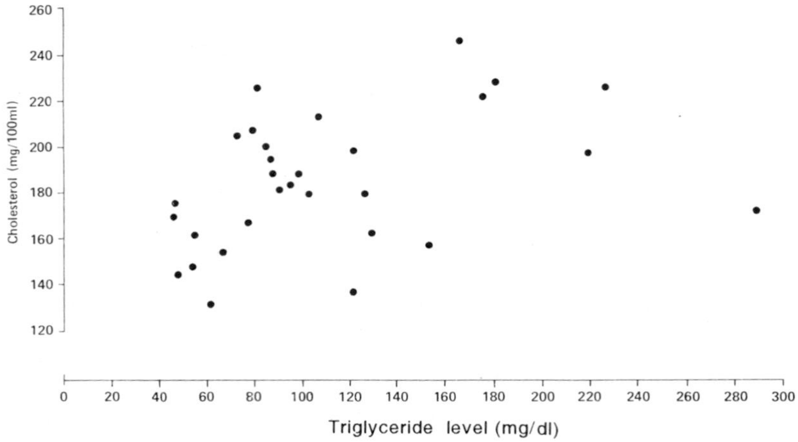

Figure 3–7. Scatter plot of cholesterol versus triglyceride levels of 30 male medical students.

Many other types of graphs are possible, but only two more will be mentioned here. The first, the **scatter plot,** or **scatter diagram,** is an effective way of illustrating the relationship between two measures. In it every point represents a pair of values, such as the values of two different measures taken on the same person. Thus, in the scatter plot depicted in Figure 3–7, every point represents a triglyceride level taken from Table 3–1, together with a corresponding cholesterol level measured on the same blood sample. We can see that there is a slight tendency for one measure to depend on the other, a fact that would not have been clear if we had simply listed each cholesterol level together with the corresponding triglyceride level.

The final graph that we shall mention here is the **tree diagram.** This is often used to help in making decisions, in which case it is called a **decision tree.** A tree diagram displays in temporal sequence possible types of actions or outcomes. Figure 3–8 gives a very simple example; it indicates the possible outcomes and their relative frequencies following myocardial infarction. This kind of display is often much more effective than a verbal description of the same information. Tree diagrams are also often helpful in solving problems.

PROPORTIONS AND RATES

In comparing the number of events occurring in two groups, raw numbers are difficult to interpret unless each group contains the same number

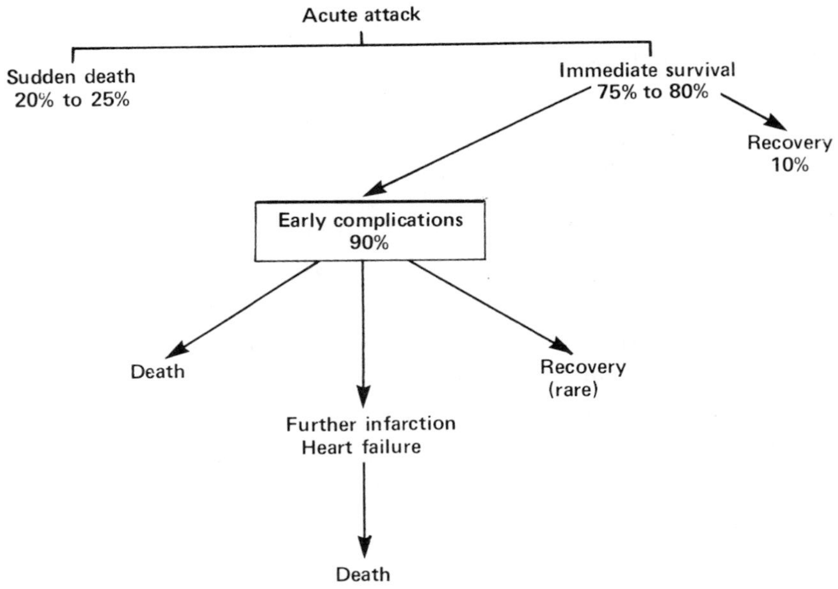

Figure 3–8. Tree diagram indicating outcome of myocardial infarction. (Source: R.A. Cawson, A.W. McCracken and P.B. Marcus (1982): Pathologic mechanisms and human disease. C.V. Mosby Co.)

of persons. We often compute **proportions** or **percentages** to facilitate such comparisons. Thus, if the purpose of a measure is to determine whether the inhabitants in one community have a more frequent occurrence of tuberculosis than those in another, simple counts have obvious shortcomings. Community A may have more cases than Community B because its population is larger. To make a comparison, we need to know the proportionate number of cases in each community. Again, it may be necessary to specify the time at or during which the events of interest occur. Thus, if 500 new cases of tuberculosis were observed in a city of 2 million persons in 1973, we say that 0.025% of the population in this city developed tuberculosis in 1973.

Sometimes it is more convenient to express proportions multiplied by some number other than 100 (which results in a percentage). Thus, the new cases of tuberculosis in a city for the year 1993 might be expressed as 500 cases per 2 million persons (the actual population of the city), 0.025 per hundred (percent), 0.25 per thousand, or 250 per million. We see that three components are required for expressions of this type:

1. The number of individuals in whom the disease, abnormality, or other characteristic occurrence is observed (the numerator).

2. The number of individuals in the population among whom the characteristic occurrence is ascertained (the denominator).

3. A specified period of time during which the disease, abnormality, or characteristic occurrence is observed.

The numerator and the denominator should be similarly restricted; if the numerator represents a count of persons who have a characteristic in a particular age-race-sex group, then the denominator should also pertain to that same age-race-sex group. When the denominator is restricted solely to those persons who are capable of having or contracting a disease, it is sometimes referred to as a population at risk. For example, a hospital may express its maternal mortality as the number of maternal deaths per thousand deliveries. The women who delivered make up the population at risk for maternal deaths. Similarly, case fatality is the number of deaths due to a disease per so many patients with the disease; here the individuals with the disease constitute the population.

All such expressions are just conversions of counts into proportions, or fractions of a group, in order to summarize data so that comparisons can be made among groups. They are commonly called **rates,** though strictly speaking a rate is a measure of the rapidity of change of a phenomenon, usually per unit time. Expressions such as "maternal death rate" and "case fatality rate" are often used to describe these proportions even when no concept of a rate per unit time is involved. One of the main concerns of epidemiology is to find and enumerate appropriate denominators for describing and comparing groups in a meaningful and useful way. Two other commonly seen but often confused measures of disease frequency used in epidemiology are prevalence and incidence. The **prevalence** of a disease is the number of cases (of that disease) at a given point in time. Prevalence is usually measured as the ratio of the number of cases at a given point in time to the number of people in the population of interest at that point in time.

The **incidence** of a disease is the number of new cases that occur during a specified amount of time. To adjust for the size of the population being observed, incidence is usually measured as the ratio of the number of new cases occurring during a period to the number of people initially at risk of developing the disease. Thus, if the number of cases of coronary heart disease events (such as angina, myocardial infarction, or sudden death) occurring in a population of 742 men during a 12-year period is observed to be 88, the incidence rate would be computed as follows:

$$I = \frac{88}{742} \times 1{,}000 = 119$$

Thus, the incidence is 119 coronary-heart-disease events per 1,000 persons initially at risk, during the 12-year period. If this were expressed *per year*, it would be a true rate.

Figure 3–9 demonstrates the difference between incidence and prevalence. Assume that each line segment represents a case of disease from time of onset (beginning of the line segment) until the disease has run its course (end of the line segment). Moreover, assume that 1,000 persons are at risk on any given day. The incidence for day 1 is 4 cases per 1,000 persons (4 new line segments) and for day 2 it is 2 cases per 1,000 persons (2 new line segments). The prevalence at the end of day 1 is 4 per 1,000 (4 line segments exist), and at the end of day 2 it is 6 (6 line segments exist). It should be obvious that two diseases can have identical incidence, and yet one would have a much higher prevalence if its duration (time from onset until the disease has run its course) is much larger.

Incidence measures rate of development of disease. It is therefore a measure of risk of disease and is useful in studying possible reasons (or causes) for disease developing. We often study the incidence in different groups of people and then try to determine the reasons why it may be higher in one of the groups. Prevalence measures the amount of disease in a population at a given point in time. Because prevalence is a function of the duration of a disease, it is of more use in planning health-care services for that disease.

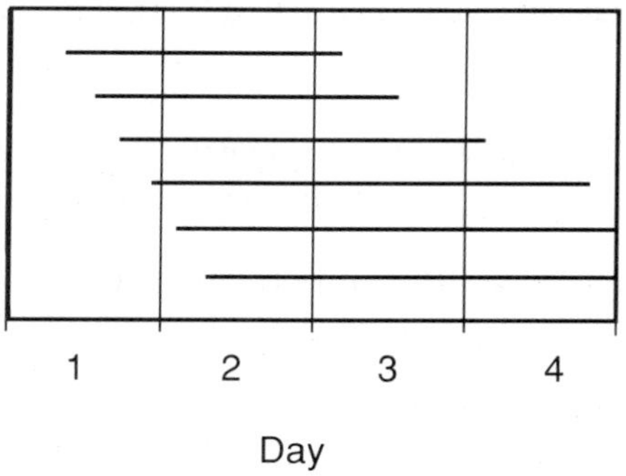

Figure 3–9. Six cases of a disease represented over time by line segments.

RELATIVE MEASURES OF DISEASE FREQUENCY

Several methods have been developed for measuring the relative amount of disease occurring in different populations. For example, we might wish to measure the amount of disease occurring in a group exposed to some environmental condition, such as cigarette smoking, relative to that in a group not exposed to that condition. One measure used for this purpose is the **relative risk** (RR), which is defined as

$$\text{RR} = \frac{\text{incidence of disease in exposed group}}{\text{incidence of disease in unexposed group}}$$

For example, if the incidence of a particular disease in a group exposed to some condition is 30 per 100,000 per year, compared with an incidence of 10 per 100,000 per year in a group unexposed to the condition, then the relative risk (exposed versus unexposed) is

$$\text{RR} = \frac{\text{30 per 100,000 per year}}{\text{10 per 100,000 per year}} = 3$$

The phrase "exposed to a condition" is used in a very general sense. Thus one can talk of the relative risk of ankylosing spondylitis to a person possessing the HLA antigen B27, versus not possessing that antigen, though of course the antigen is inherited from a parent rather than acquired from some kind of environmental exposure.

Another relative measure of disease occurrence is the **odds ratio.** The odds in favor of a particular event are defined as the frequency with which the event occurs divided by the frequency with which it does not occur. For a disease with an incidence of 30 per 100,000 per year, for example, the odds in favor of the disease are 30/99,970. The odds ratio (OR) is then defined as

$$\text{OR} = \frac{\text{odds in favor of disease in exposed group}}{\text{odds in favor of disease in unexposed group}}$$

Thus, if the incidences are 30 per 100,00 and 10 per 100,000 as above, the odds ratio for exposed versus unexposed is

$$\text{OR} = \frac{30}{99{,}970} \bigg/ \frac{10}{99{,}990} = 3.00006$$

You can see from this example that for rare diseases the odds ratio closely approximates the relative risk. The attractive feature of the odds ratio is that it can be estimated without actually knowing the incidences. This is often done in case-control studies, which were described in Chapter 2. Suppose, for example, it is found that 300 out of 1,000 cases of a disease had previous exposure to a particular condition, whereas only 100 out of 1,000 controls were similarly exposed. These data tell us nothing about the incidences of the disease among exposed and unexposed persons, but they do allow us to calculate the odds ratio, which in this case is

$$\frac{300}{700} \bigg/ \frac{100}{900} \cong 4$$

The last relative measure of disease frequency we shall discuss is the **attributable risk**, defined as the incidence of disease in an exposed group minus the incidence of disease in an unexposed group. Thus, in the previous example, the attributable risk (AR) is

$$\text{AR} = 30 - 10 = 20 \text{ per } 100{,}000 \text{ per year}$$

An excess of 20 cases per 100,000 per year can be attributed to exposure to the particular condition. Sometimes we express attributable risk as a percentage of the incidence of disease in the unexposed group. In the above example, we would have

$$\text{AR\%} = \frac{30 - 10}{10} \times 100 = 200\%$$

In this case we could say there is a 200% excess risk of disease in the exposed group.

SENSITIVITY, SPECIFICITY, AND PREDICTIVE VALUES

We now define some terms that are often used to measure the effectiveness of a test procedure, such as a laboratory test to help diagnose a disease. We shall illustrate these terms using the following hypothetical population of 10,000 persons classified on the basis of disease status and their reaction to the test:

Disease status	Test Result: Negative	Test Result: Positive	Total
Absent	8,820	980	9,800
Present	20	180	200
Total	8,840	1,160	10,000

Note first that the prevalence of the disease in the population is 200/10,000, or 2%.

The **sensitivity** of the test measures how well it detects disease; it is the proportion of those with the disease who give a positive result. In the example the sensitivity is 180/200 = 0.9.

The **specificity** of the test measures how well it detects absence of disease; it is the proportion of those without the disease who give a negative result. In the example, the specificity is 8,820/9,800 = 0.9.

Whenever sensitivity and specificity are equal, they represent the proportion of the population that is correctly classified by the test. Thus, in our example, 90% of the total population is correctly classified by the test. This does not mean, however, that 90% of those who give a positive result have the disease. In order to know how to interpret a particular test result, we need to know the **predictive values** of the test, which are defined as the proportion of those positive who have the disease, and the proportion of those negative who do not have the disease. For our example, these values are 180/1,160 = 0.155, and 8,820/8,840 = 0.998, respectively. Especially in the case of a rare disease, a high specificity and high sensitivity are not sufficient to ensure that a large proportion of those who test positive actually have the disease.

MEASURES OF CENTRAL TENDENCY

Measures of central tendency, or measures of location, tell us where on our scale of measurement a set of values tends to lie. All the values in Table 3–1, for example, lie between 45 and 287 mg/dl, and we need our measure of central tendency to be somewhere between these two values. If our values had been in mg per liter, on the other hand, we should want our measure of central tendency to be 10 times as large. We shall discuss three measures of central tendency: the mean, the median, and the mode.

They all have the property (when used to describe continuous data) that if every value in our dataset is multiplied by a constant number, then the measure of central tendency is multiplied by the same number. Similarly, if a constant is added to every value, then the measure of central tendency is increased by that same amount.

The **mean** of a set of numbers is just their numerical average. You know, for example, that to compute your mean score for four test grades you add the grades and divide by four. If your grades were 94, 95, 97, and 98, your mean score would be $(94 + 95 + 97 + 98)/4 = 384/4 = 96$.

One of the disadvantages of the mean as a summary statistic is that it is sensitive to unusual values. The mean of the numbers 16, 18, 20, 22 and 24 is 20, and indeed 20 appears to be typical of these numbers. The mean of the numbers 1, 2, 3, 4 and 90 is also 20, but 20 is not typical of these numbers because of the one unusual value. Another disadvantage of the mean is that strictly speaking it should be used only for data measured on an interval scale, because implicit in its use is the assumption that the units of the scale are all of equal value. The difference between 50 and 51 mg/dl of triglyceride is in fact the same as the difference between 250 and 251 mg/dl of triglyceride, (i.e., 1 mg/dl). Because of this, it is meaningful to say that the mean of the 30 values in Table 3–1 is 111.2 mg/dl. But if the 30 students had been scored on an 11-point scale, 0 through 10 (whether for triglyceride level or anything else), the mean score would be strictly appropriate only if each of the 10 intervals, 0 to 1, 1 to 2, etc., were equal in value. Nevertheless, the mean is the most frequently used descriptive statistic because, as we shall see later, it has statistical properties that make it very advantageous if no unusual values are present.

The **median** is the middlemost value in a set of ranked data. Thus the median of the numbers 16, 18, 20, 22, and 24 is 20. The median of the numbers 1, 2, 3, 4, and 90 is 3. In both sets of numbers the median is typical of most of the data, so the median has the advantage of not being sensitive to unusual values. If the set of data contains an even number of values, then the median lies between the two middlemost values, and we usually just take their average. Thus the median of the data in Table 3–1 lies between 90 and 93 mg/dl, and we would usually say the median is 91.5 mg/dl.

A percentile is the value of a trait at or below which the corresponding percentage of a dataset lies. If your grade on an examination is at the 90th percentile, then 90% of those taking the examination obtained the same or a lower grade. The median is thus the 50th percentile—the point at or below which 50% of the data points lie. The median is a proper measure of central tendency for data measured either on an interval or on an ordinal scale, but cannot be used for nominal data.

The **mode** is defined as the most frequently occurring value in a set of data. Thus, for the data 18, 19, 21, 21, 22, the value 21 occurs twice, whereas all the other values occur only once, and so 21 is the mode. In the case of continuous data, the mode is related to the concept of a peak in the frequency distribution. If there is only one peak, the distribution is said to be unimodal; if there are two peaks, it is said to be bimodal, etc. Thus the distribution depicted in Figure 3–1 is unimodal, and the mode is clearly between 180 and 190 mg/dl.

An advantage of the mode is that it can be used for nominal data: the modal category is simply the category that occurs most frequently. But it is often difficult to use for a small sample of continuous data. What, for example, is the mode of the data in Table 3–1? Each value occurs exactly once, so shall we say there is no mode? The data can be grouped as in Figure 3–2, and then it appears that the 70- to 90-mg/dl category is the most frequent. But with this grouping we also see peaks (and hence modes) at 150 to 190, 210 to 230, and 270 to 290 mg/dl. For this reason the mode is less frequently used as a measure of central tendency in the case of continuous data.

MEASURES OF SPREAD OR VARIABILITY

Suppose you score 80% on an examination and the average for the class is 87%. Suppose you are also told that the grades ranged from 79% to 95%. Obviously you would feel much better had you been told that the spread was from 71% to 99%. The point here is that it is often not sufficient to know the mean of a set of data, but, rather, it is of interest to know the mean together with some **measure of spread** or **variability.**

The **range** is the largest value minus the smallest value. It provides a simple measure of variability but is very sensitive to one or two extreme values. The range of the data in Table 3–1 is 287 − 45 = 242 mg/dl, but it would be only 173 mg/dl if the two largest values were missing. Percentile ranges are less sensitive and provide a useful measure of dispersion in data. For example, the 90th percentile minus the 10th percentile, or the 75th percentile minus the 25th percentile, can be used. The latter is called the **interquartile range.** For the data in Table 3–1 the interquartile range is 124 − 67 = 57 mg/dl. (For 30 values we cannot obtain the 75th and 25th percentiles accurately, so we take the next lowest percentiles: 124 is the 22nd out of 30 values, or 73rd percentile, and 67 is the 7th out of 30, or 23rd percentile.) If the two largest values were missing from the table, the interquartile range would be 123 − 67 = 56 mg/dl, i.e., almost the same as for all 30 values.

The variance or its square root, the standard deviation, is perhaps the most frequently used measure of variability. The variance, denoted s^2, is basically the average squared deviation from the mean. To compute the variance of a set of data, we

1. Subtract the mean from each value to get a "deviation" from the mean.

2. Square each deviation from the mean.

3. Sum the squares of the deviations from the mean.

4. Divide the sum of squares by one less than the number of values in the set of data.

Thus, for the numbers 18, 19, 20, 21, and 22, we find that the mean is (18 + 19 + 20 + 21 + 22)/5 = 20, and the variance is computed as follows:

1. Subtract the mean from each value to get a deviation from the mean, which we shall call d:

	d
18 − 20 =	−2
19 − 20 =	−1
20 − 20 =	0
21 − 20 =	+1
22 − 20 =	+2

2. Square each deviation, d, to get squares of deviations, d^2:

d	d^2
−2	4
−1	1
0	0
+1	1
+2	4

3. Sum the squares of the deviations:

$$4 + 1 + 0 + 1 + 4 = 10$$

4. Divide the sum of squares by one less than the number of values in the set of data:

$$\text{Variance} = s^2 = \frac{10}{5-1} = \frac{10}{4} = 2.5$$

The standard deviation is just the square root of the variance; that is, in this example,

$$\text{Standard deviation} = s = \sqrt{2.5} = 1.6$$

Notice that the variance is expressed in squared units, whereas the standard deviation gives results in terms of the original units. If, for example, the original units for the above data were years (e.g., years of age), then s^2 would be 2.5 $(\text{years})^2$, and s would be 1.6 years.

As a second example, suppose the numbers were 1, 2, 3, 4, and 90. Again, the average is $(1 + 2 + 3 + 4 + 90)/5 = 20$, but the data are quite different from those in the previous example. Here,

$$s^2 = \frac{(1-20)^2 + (2-20)^2 + (3-20)^2 + (4-20)^2 + (90-20)^2}{4}$$

$$= \frac{(-19)^2 + (-18)^2 + (-17)^2 + (-16)^2 + (70)^2}{4}$$

$$= \frac{6{,}130}{4} = 1{,}532.5$$

The standard deviation is $s = \sqrt{1{,}532.5} = 39.15$. Thus, you can see that the variance and the standard deviation are larger for a set of data that is obviously more variable.

Two questions you may have concerning the variance and the standard deviation are: Why do we square the deviations, and why do we divide by one less than the number of values in the set of data being considered? Look back at step 2 and see what would happen if we did not square the deviations, but simply added up the unsquared deviations d. Because of the way in which the mean is defined, the deviations always add up to zero! Squaring is a simple device to stop this happening. When we average the squared deviations, however, we divide by one less than the number of values in the data set. The reason for this is that it leads to an unbiased estimate, a concept we shall explain more fully in Chapter 6. For the moment, just note that if the data set consisted of an infinite number of values (which is conceptually possible for a whole population), it would make no difference whether or not we subtracted one from the divisor.

The last measure of spread we shall discuss is the **coefficient of variation.** This is the standard deviation expressed as a proportion or percentage of the mean. It is a dimensionless measure and, as such, it is a useful descriptive index for comparing the relative variability in two sets of values. Suppose, for example, that we wished to compare the variability of birth weight with the variability of adult weight. Clearly, on an absolute scale, birth weights must vary much less than adult weights simply because they are necessarily limited to being much smaller. As a more extreme example, suppose we wished to compare the variability in weights of ants and elephants! In such a situation it makes more sense to express variability on a relative scale. Thus, we can make a meaningful comparison of the variability in two sets of numbers with different means by observing the difference between their coefficients of variation. As an example, suppose the mean of a set of cholesterol levels is 219 mg/dl and the standard deviation is 14.3 mg/dl. The coefficient of variation, as a percentage, is then

$$\text{CV\%} = \frac{\text{standard deviation}}{\text{mean}} \times 100$$

$$= \frac{14.3\text{mg/dl}}{219\text{ mg/dl}} \times 100$$

$$= 6.5$$

This could then be compared with, for example, the coefficient of variation of triglyceride levels.

MEASURES OF SHAPE

There are many other descriptive statistics, some of which will be mentioned in later chapters of this book. We shall conclude this chapter with the names of a few statistics that describe the shape of distributions. (Formulas for calculating these statistics, as well as others, are presented in Appendix 1).

The **coefficient of skewness** is a measure of symmetry. A symmetric distribution has a coefficient of skewness that is zero. As illustrated in

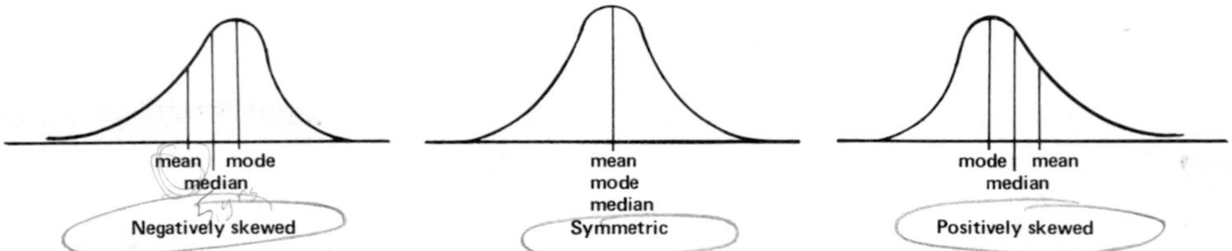

Figure 3–10. Examples of negatively skewed, symmetric, and positively skewed distributions.

Figure 3–10, a distribution that has a tail to the left has a negative coefficient of skewness and is said to be negatively skewed; one that has a tail to the right has a positive coefficient of skewness and is said to be positively skewed. Note that in a symmetric unimodal distribution the mean, the median, and the mode are all equal. In a unimodal asymmetric distribution, the median always lies between the mean and the mode. The serum triglyceride values in Table 3–1 have a positive coefficient of skewness, as can be seen in the histogram in Figure 3–2.

The **coefficient of kurtosis** measures the peakedness of a distribution. In Chapter 6 we shall discuss a very important distribution, called the normal distribution, for which the coefficient of kurtosis is 3. A distribution with a larger coefficient than this is leptokurtic ("lepto" means slender), and one with a coefficient smaller than this is platykurtic ("platy" means flat or broad). Kurtosis, or peakedness, is also often measured by the standardized "fourth cumulant" (denoted g_2), which is the coefficient of kurtosis minus three; on this scale, the normal distribution has zero kurtosis. Different degrees of kurtosis are illustrated in Figure 3–11.

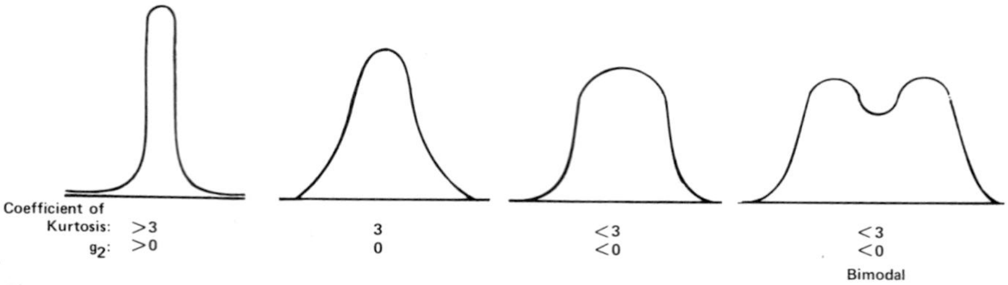

Figure 3–11. Examples of symmetric distributions with coefficient of kurtosis greater than 3 ($g_2 > 0$), equal to 3 ($g_2 = 0$, as for a normal distribution), and less than 3 ($g_2 < 0$).

SUMMARY

1. Continuous data arise only from quantitative traits, whereas categorical or discrete data arise either from quantitative or from qualitative

traits. Continuous data are measured on an interval scale, categorical data on either an ordinal (that can be ordered) or a nominal (name only) scale.

2. Descriptive statistics, tables, and graphs summarize the essential characteristics of a set of data.

3. A table should be easy to read. The title should indicate what is being tabulated, with the units of measure.

4. Bar graphs are used for discrete data, histograms and frequency polygons for continuous data. A cumulative plot has the advantage that every data point can be represented in it. A scatter plot or scatter diagram illustrates the relationship between two measures. A tree diagram displays a sequence of actions and/or results.

5. Proportions and rates allow one to compare counts when denominators are appropriately chosen. The term "rate" properly indicates a measure of rapidity of change but is often used to indicate a proportion multiplied by some number other than 100. Prevalence is the number or proportion of cases present at a particular time; incidence is the number or proportion of new cases occurring in a specified period.

6. Relative risk is the incidence of disease in a group exposed to a particular condition, divided by the incidence in a group not so exposed. The odds ratio is the ratio of the odds in favor of a disease in an exposed group to the odds in an unexposed group. In the case of a rare disease, the relative risk and the odds ratio are almost equal. Attributable risk is the incidence of a disease in a group with a particular condition minus the incidence in a group without the condition, often expressed as a percentage of the latter.

7. The sensitivity of a test is the proportion of those with the disease who give a positive result. The specificity of a test is the proportion of those without the disease who give a negative result. In the case of a rare disease, it is quite possible for the test to have a low predictive value even though these are both high. The predictive values are defined as the proportion of the positives who have the disease and the proportion of the negatives who do not have the disease.

8. Three measures of central tendency, or location, are the mean (arithmetic average), the median (50th percentile), and the mode (one or

more peak values). All three are equal in a unimodal symmetric distribution. In a unimodal asymmetric distribution, the median lies between the mean and the mode.

9. Three measures of spread, or variability, are the range (largest value minus the smallest value), the interquartile range (75th percentile minus the 25th percentile), and the standard deviation (square root of the variance). The variance is basically the average squared deviation from the mean, but the divisor used to obtain this average is one less than the number of values being averaged. The variance is expressed in squared units. The coefficient of variation, which is dimensionless, is the standard deviation divided by the mean (and multiplied by 100 if it is expressed as a percentage).

10. An asymmetric distribution may be positively skewed (tail to the right) or negatively skewed (tail to the left). A distribution may be leptokurtic (peaked) or platykurtic (flat-topped or multimodal).

FURTHER READING

Elandt-Johnson, R.C. (1975) Definition of Rates: Some Remarks on Their Use and Misuse. American Journal of Epidemiology 102:267–271. (This gives very precise definitions of ratios, proportions, and rates; a complete understanding of this paper requires some mathematical sophistication.)

Stevens, S.S. (1946) On the Theory of Scales of Measurement. Science 103:677–680. (This article defines in more detail four hierarchical categories of scales for measurements—nominal, ordinal, interval, and ratio.)

Wainer, H. (1984) How to Display Data Badly. The American Statistician 38:137–147. (Though pointed in the wrong direction, this is a serious article. It illustrates the 12 most powerful methods—the dirty dozen— of misusing graphics.)

PROBLEMS

1. A nominal scale is used for
   - A. all categorical data
   - B. discrete data with categories that do not follow a natural sequence
   - C. continuous data that follow a natural sequence
   - D. discrete data with categories that follow a natural sequence
   - E. quantitative data

2. The following are average annual incidences per 1 million for testicular cancers, New Orleans, 1974 to 1977:

Age	White	Black	Odds Ratio
15–19	29.4	13.4	2.2
20–29	113.6	9.5	12.0
30–39	91.0	49.8	1.8
40–49	75.5	0.0	—
50–59	50.2	22.2	2.3
60–69	0.0	0.0	—
70+	38.2	0.0	—

Based on these data, which of the following is true of New Orleans males?

A. There is no difference in the risk of developing testicular cancer for blacks and whites.
B. The odds of developing testicular cancer are greater in blacks than in whites.
C. The racial difference in risk of developing testicular cancer cannot be determined from these data.
D. The risk of developing testicular cancer is greater in whites than in blacks in virtually every age group.

3. Refer to the diagram below. Each horizontal line in the diagram indicates the month of onset and the month of termination for one of 24 episodes of disease. Assume an exposed population of 1,000 individuals in each month.

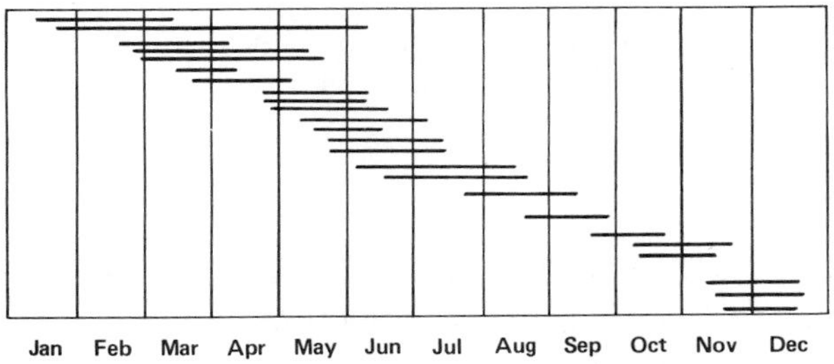

(1) The incidence for this disease during April was

A. 2 per 1,000
B. 3 per 1,000
C. 6 per 1,000
D. 7 per 1,000
E. 9 per 1,000

(2) The prevalence on March 31 was

A. 2 per 1,000
B. 3 per 1,000

C. 6 per 1,000
D. 7 per 1,000
E. 9 per 1,000

4. The incidence of a certain disease during 1987 was 16 per 100,000 persons. This means that for every 100,000 persons in the population of interest, 16 people
   A. had the disease on January 1, 1987
   B. had the disease on December 31, 1987
   C. developed the disease during 1987
   D. developed the disease each month during 1987
   E. had disease with duration one month or more during 1987

5. A large study of bladder cancer and cigarette smoking produced the following data:

	Incidence of Bladder Cancer (per 100,000 males per year)
Cigarette smokers	48.0
Nonsmokers	25.4

   The relative risk of developing bladder cancer for male cigarette smokers compared with male nonsmokers is
   A. $48.0/25.4 = 1.89$
   B. unknown
   C. $48.0 - 25.4 = 22.6$
   D. 48.0
   E. $\frac{48.0 - 25.4}{48.0}$

6. Both the specificity and sensitivity of a diagnostic test for a particular disease are 0.99. All the following are necessarily true except
   A. a person who is positive for the test has a 99% chance of having the disease
   B. a person without the disease has a 99% chance of being negative for the test
   C. a person has a 99% chance of being correctly classified by the test
   D. a person with the disease has a 99% chance of being positive for the test

7. The specificity of a test is reported as being 0.80. This means that
   A. the test gives the correct result in 80% of persons tested
   B. disease is present in 80% of persons who test positive
   C. disease is absent in 80% of persons who test negative

D. the test is positive in 80% of persons tested who have the disease
E. the test is negative in 80% of persons tested who are disease-free

8. Most values in a small set of data range from 0 to 35. The data are highly skewed, however, with a few values as large as 55 to 60. The best measure of central tendency is the
   A. mean
   B. median
   C. mode
   D. standard deviation
   E. range

9. One useful summary of a set of data is provided by the mean and standard deviation. Which of the following is true?
   A. The mean is the middle value (50th percentile) and the standard deviation is the difference between the 90th and the 10th percentiles.
   B. The mean is the arithmetic average and the standard deviation measures the extent to which observations vary about or are different from the mean.
   C. The mean is the most frequently occurring observation and the standard deviation measures the length of a deviation.
   D. The mean is half the sum of the largest and smallest value and the standard deviation is the difference between the largest and smallest observations.
   E. None of the above.

10. All of the following are measures of spread except
   A. variance
   B. range
   C. mode
   D. standard deviation
   E. coefficient of variation

11. The height in centimeters of second-year medical students was recorded. The variance of these heights was calculated. The unit of measurement for the calculated variance is
   A. $\sqrt{\text{centimeters}}$
   B. centimeters
   C. $(\text{centimeters})^2$
   D. unit free
   E. none of the above

12. The standard deviation for Dr. A's data was found to be 10 units, while that for Dr. B's data was found to be 15 units. This suggests that Dr. A's data are
   A. larger in magnitude on average
   B. skewed to the right

C. less variable
D. unbiased
E. unimodal

13. Consider the following sets of cholesterol levels in milligrams per deciliter (mg/dl):

Set 1: 200, 210, 190, 220, 195
Set 2: 210, 170, 180, 235, 240

The standard deviation of set 1 is

A. the same as that of Set 2
B. less than that of Set 2
C. greater than that of Set 2
D. equal to the mean for Set 2
E. indeterminable from these data

14. The following is a histogram for the pulse rates of 1,000 students:

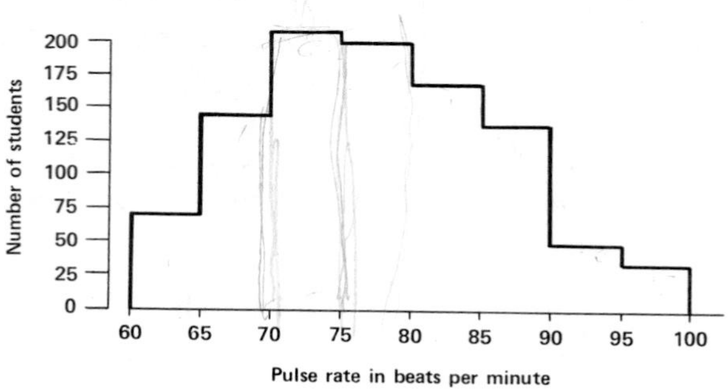

Which of the following is between 70 and 75 beats per minute?

A. the mode of the distribution
B. the median of the distribution
C. the mean of the distribution
D. the range of the distribution
E. none of the above

mean = average
median = middle most #
mode = most frequent

15. The following cumulative plot was derived from the pulse rates of 1,000 students:

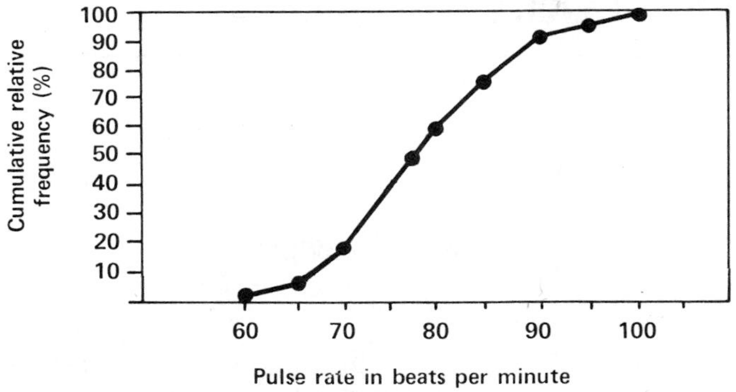

Which of the following is false?

A. the range of the distribution is 60 to 100 beats per minute
B. the mode of the distribution is 100 beats per minute
C. the median of the distribution is 77 beats per minute
D. 92% of the values are less than 90 beats per minute
E. 94% of the values are greater than 65 beats per minute

CHAPTER FOUR

Key Concepts

three definitions of probability
the probability of either of two events, the joint probability of two events, conditional probability, mutually exclusive events, independent events
Bayes' theorem
likelihood ratio

Symbols and Abbreviations

P(A)	probability of the event A
P(A or B)	probability of either event A or event B
P(A and B)	joint probability of events A and B
P(A|B)	conditional probability of event A given event B

The Laws of Probability

DEFINITION OF PROBABILITY

Although the meaning of a statement such as, "The probability of rain today is fifty percent," may seem fairly obvious, it is not easy to give an exact definition of the term "probability." In fact, three different definitions have been proposed, each focusing on a different aspect of the concept. Since probability plays such a fundamental role in biostatistics, we start by reviewing all three of these definitions before stating the mathematical laws that govern its manipulation. The mathematicians who originally studied probability were motivated by gambling and so used games of chance (e.g., cards and dice) in their studies. It will be convenient for us to use similar examples initially.

The classical definition of probability can be stated as follows: Given a set of equally likely possible outcomes, the probability P of the event A, which we write P(A), is the number of outcomes that are "favorable to" A divided by the total number of possible outcomes:

$$P(A) = \frac{\text{number of outcomes favorable to A}}{\text{total number of possible outcomes}}$$

This definition will become clearer when we illustrate it with some examples. Suppose we pick a single card at random from a well-shuffled regular deck of 52 cards comprising four suits, with 13 cards in each suit. What is the probability that the card is an ace? We let A be the event "the card is an ace." Each card represents a possible outcome, and so the total number of possible outcomes is 52; the number of outcomes that are "favorable to" A is four, since there are four aces in the deck. Therefore the probability is

$$P(A) = \frac{4}{52} = \frac{1}{13}$$

As another example, what is the probability of obtaining two 6s when two normal six-sided dice are rolled? In this case, let A be the event "obtaining two sixes." Each of the two dice can come up one, two, three, four, five, or six, and so the total number of possible outcomes is 36 (6×6). Only one of these outcomes (six and six) is "favorable to" A, and so the probability is $P(A) = 1/36$.

This definition of probability is precise and appears to make sense. Unfortunately, however, it contains a logical flaw. Note that the definition includes the words "equally likely," which is another way of saying "equally probable." In other words, *probability has been defined in terms of probability!* Despite this difficulty, our intuition tells us that when a card is taken at random from a well-shuffled deck, or when normal dice are rolled, there is physical justification for the notion that all possible outcomes are "equally likely." When there is no such physical justification (as we shall see in a later example), the classical definition can lead us astray if we are not careful.

The frequency definition of probability supposes that we can perform many, many replications or trials of the same experiment. As the number of trials tends to infinity (i.e., to any very large number, represented mathematically by the symbol ∞), the proportion of trials in which the event A occurs tends to a fixed limit. We then define the probability of the event A as this limiting proportion. Thus, to answer the question, "What is the probability that the card is an ace?", we suppose that many trials are performed, in each of which a card is drawn at random from a well-shuffled deck of 52 cards. After each trial we record in what proportion of the cards drawn so far an ace has been drawn, and we find that, as the number of trials increases indefinitely, this proportion tends to the limiting value of $1/13$. This concept is illustrated in Figure 4–1, which represents a set of trials in which an ace was drawn at the 10th, 20th, 40th, 54th, 66th, and 80th trials.

Since one can never actually perform an infinite number of trials, this definition of probability is mathematically unsatisfying. Nevertheless, it is the best way of interpreting probability to a patient in practical situations. The statement "There is a fifty percent probability of rain today" can be interpreted to mean that on just half of many days on which such a state-

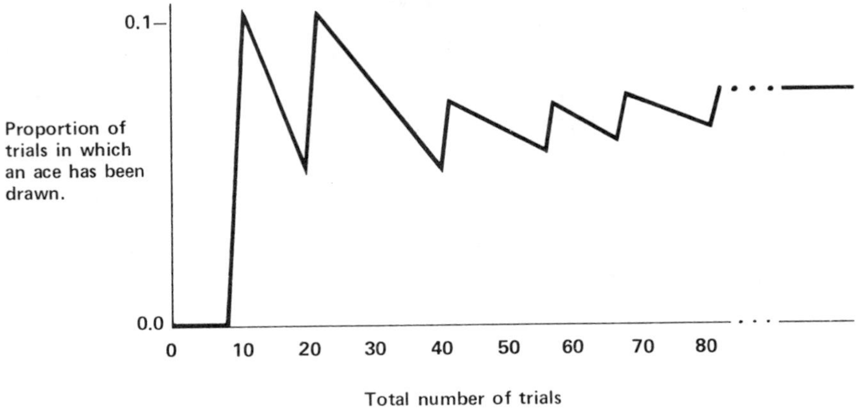

Figure 4–1. Example of probability defined as the limiting proportion of trials, as the number of trials tends to infinity, in which a particular event occurs.

ment is made, it will rain. Provided that this is true, the probability statement is valid.

In medical practice, however, quoting valid probabilities may not be sufficient. When a pregnant woman is counseled concerning the probability that her child will have a particular disease, the probability that is quoted must be both valid and relevant. Suppose, for example, a 25-year-old woman has a child with Down syndrome (trisomy 21), and, on becoming pregnant again, seeks counseling. Her physician checks published tables and finds out that the probability of a 25-year-old woman having a Down syndrome child is about 1 in 400. If she is counseled that the probability of her second child having Down syndrome is only about 1 in 400, she will have been quoted a valid probability; only 1 in 400 such women coming for counseling will bear a child with Down syndrome. But this probability will not be relevant if the first child has Down syndrome because of a translocation in the mother, i.e., because the mother has a long arm of chromosome 21 attached to another of her chromosomes. If this is the case, the risk to the unborn child is about 1 in 6. A physician who failed to recommend the additional testing required to arrive at such a relevant probability for a specific patient could face embarrassment, perhaps even a malpractice suit, though the quoted probability, on the basis of all the information available, was perfectly valid.

The mathematical (axiomatic) definition of probability avoids the disadvantages of the two other definitions and is the definition used by mathematicians. A simplified version of this definition, which, although incomplete, retains its main features, is as follows. A set of probabilities is any set of numbers for which (1) each number is positive (greater than or equal to zero), and (2) the sum of the numbers is unity (one). This definition, unlike the other two, gives no feeling for the practical meaning of probability. It does, however, describe the essential characteristics of probability: there is a set of possible outcomes, each associated with a positive probability of occurring, and at least one of these outcomes must occur. We now turn to some fundamental laws of probability, which do not depend on which definition is taken.

THE PROBABILITY OF EITHER OF TWO EVENTS: A OR B

If A and B are **two events,** what is the **probability of either A or B occurring**? Let us consider a simple example. A regular deck of cards is shuffled well and a card is drawn. What is the probability that it is either an ace or a king? The deck contains eight cards that are either aces or kings, and so the answer is $8/52 = 2/13$. Now notice that the two events "ace" and "king" are **mutually exclusive,** in that when we draw a card from the deck, it cannot be both an ace and a king. **If A and B are mutually exclusive** events, then

$$P(\text{A or B}) = P(\text{A}) + P(\text{B})$$

mutually exclusive

Thus, in our example, we have

$$\begin{aligned} P(\text{ace or king}) &= P(\text{ace}) + P(\text{king}) \\ &= \frac{1}{13} + \frac{1}{13} \\ &= \frac{2}{13} \end{aligned}$$

Now suppose the question had been: "What is the probability that the card is an ace or a heart?" In this case the events ace and "heart" are *not* mutually exclusive, because the same card could be both an ace and a heart; therefore we cannot use the formula given above. How many cards in the deck are either aces or hearts? There are 4 aces and 13 hearts, but a total of only 16 cards that are either an ace or a heart; the ace of clubs, the ace of diamonds, the ace of spades, and the 13 hearts. Notice that if we carelessly write "4 aces + 13 hearts = 17 cards," the ace of hearts has been counted twice—once as an ace and once as a heart. In other words, the number of cards that are either aces or hearts is the number of aces, plus the number of hearts, minus the number of cards that are both aces and hearts (one, in this example). Analogously, dividing each of these numbers by 52 (the total number of cards in the deck), we have

$$
\begin{aligned}
\text{P(ace or heart)} &= \text{P(ace)} + \text{P(heart)} - \text{P(ace and heart)} \\
&= \frac{4}{52} + \frac{13}{52} - \frac{1}{52} \\
&= \frac{16}{52} = \frac{4}{13}
\end{aligned}
$$

The general rule for any two events A and B is

$$\text{P(A or B)} = \text{P(A)} + \text{P(B)} - \text{P(A and B)}$$

This rule is *general,* by which we mean it is *always* true. (A layman uses the word "generally" to mean "usually." In mathematics a *general* result is one that is always true, and the word "generally" means "always.")

If the event A is "the card is a king" and the event B is "the card is an ace," we have

$$
\begin{aligned}
\text{P(A or B)} &= \text{P(A)} + \text{P(B)} - \text{P(A and B)} \\
&= \frac{1}{13} + \frac{1}{13} - 0 \\
&= \frac{2}{13} \quad \text{as before}
\end{aligned}
$$

In this case, P(A and B) = 0 because a card cannot be both a king and an ace. In the special case in which A and B are mutually exclusive events, P(A and B) = 0.

THE JOINT PROBABILITY OF TWO EVENTS: A AND B

We have just seen that the probability that both events A and B occur is written P(A and B); this is also sometimes abbreviated to P(AB) or P(A, B). It is often called the **joint probability** of A and B. If A and B are mutually exclusive (i.e., they cannot both occur—if a single card is drawn, for example, it cannot be both an ace and a king), then their joint probability, P(A and B), is zero. What can we say about P(A and B) in general? One answer to this question is implicit in the general formula for P(A or B) just given. Rearranging this formula we find

$$P(A \text{ and } B) = P(A) + P(B) - P(A \text{ or } B)$$

A more useful expression, however, uses the notion of **conditional probability:** the probability of an event occurring given that another event has already occurred. We write the *conditional probability of B occurring given that A has occurred* as **P(B|A)**. Read the vertical line as "given," so that P(B|A) is the "probability of B given A" (i.e., given that A has already occurred). Sensitivity and specificity are examples of conditional probabilities. We have

$$\text{Sensitivity} = P(\text{test positive} \mid \text{disease present})$$
$$\text{Specificity} = P(\text{test negative} \mid \text{disease absent})$$

Using this concept of conditional probability, we have the following general rule for the joint probability of A and B:

$$P(A \text{ and } B) = P(A)\,P(B \mid A)$$

Since it is arbitrary which event we call A and which B, note that we could equally well have written

$$P(A \text{ and } B) = P(B)\,P(A \mid B)$$

Using this formula, what is the probability, on drawing a single card from a well-shuffled deck, that it is both an ace and a heart (i.e., that it is the ace of hearts)? Let A be the event that the card is an ace and B be the event that it is a heart. Then, from the formula:

$$P(\text{ace and heart}) = P(\text{ace})\,P(\text{heart} \mid \text{ace}) = \frac{1}{13} \times \frac{1}{4} = \frac{1}{52}$$

P(heart|ace) = ¼, because the ace that we have picked can be any of the four suits, only one of which is hearts. Similarly we could have found

$$P(\text{ace and heart}) = P(\text{heart})\,P(\text{ace} \mid \text{heart}) = \frac{1}{4} \times \frac{1}{13} = \frac{1}{52}$$

Now notice that in this particular example we have

$$P(\text{heart}\,|\,\text{ace}) = P(\text{heart}) = \frac{1}{4}$$

$$\text{and } P(\text{ace}\,|\,\text{heart}) = P(\text{ace}) = \frac{1}{13}$$

In other words, the probability of picking a heart is the same ($\frac{1}{4}$) whether or not an ace has been picked, and the probability of picking an ace is the same ($\frac{1}{13}$) whether or not a heart has been picked. These two events are therefore said to be **independent.**

Two events A and B are independent if $P(A) = P(A|B)$, or if $P(B) = P(B|A)$. From the general formula for P(A and B), it follows that if two events A and B are independent, then

$$P(A \text{ and } B) = P(A)\,P(B)$$

Conversely, two events A and B are independent if we know that

$$P(A \text{ and } B) = P(A)\,P(B)$$

It is often intuitively obvious when two events are independent. Suppose we have two regular decks of cards and randomly draw one card from each. What is the probability that the card from the first deck is a king and the card from the second deck is an ace? The two draws are clearly independent, since the outcome of the first draw cannot in any way affect the outcome of the second draw. The probability is thus

$$\frac{1}{13} \times \frac{1}{13} = \frac{1}{169}$$

But suppose we have only one deck of cards, from which we draw two cards consecutively. Now what is the probability that the first is a king and the second is an ace? Using the general formula for the joint probability of two events A and B, we have

$$\begin{aligned} P(\text{1st is king and 2nd is ace}) &= P(\text{1st is king})\,P(\text{2nd is ace}\,|\,\text{1st is king}) \\ &= \frac{4}{52} \times \frac{4}{51} = \frac{4}{663} \end{aligned}$$

In this case the two draws are not independent. The probability that the second card is an ace depends on what the first card is (if the first card is an ace, for example, the probability that the second card is an ace becomes $\frac{3}{51}$).

EXAMPLES OF INDEPENDENCE, NONINDEPENDENCE, AND GENETIC COUNSELING

It is a common mistake to assume two events are independent when they are not. Suppose two diseases occur in a population and it is known that it is impossible for a person to have both diseases. There is a strong temptation to consider two such diseases to be occurring independently in the population, whereas in fact this is impossible. Can you see why this is so? [*Hint:* Let the occurrence of one disease in a particular individual be the event A, and the occurrence of the other disease be event B; what do you know about the joint probability of A and B if (1) they are independent, and (2) they are mutually exclusive? Can P(A and B) be equal to both P(A)P(B) and zero if P(A) and P(B) are both nonzero (we are told both diseases actually occur in the population)?] On the other hand, it is sometimes difficult for an individual to believe that truly independent events occur in the manner in which they do occur. The mother of a child with a genetic anomaly may be properly counseled that she has a 25% probability of having a similarly affected child at each conception and that all conceptions are independent. But she will be apt to disbelieve the truth of such a statement if (as will happen to one quarter of mothers in this predicament, assuming the counseling is valid and she has another child) her very next child is affected.

NONINDEPENDENCE DUE TO MULTIPLE ALLELISM

Among the Caucasian population, 44% have red blood cells that are agglutinated by an antiserum denoted anti-A, and 56% do not. Similarly, the red blood cells of 14% of the population are agglutinated by another antiserum denoted anti-B, and those of 86% are not. If these two traits are distributed independently in the population, what proportion would be expected to have red cells that are agglutinated by both anti-A and anti-B? Let A+ be the event that a person's red cells are agglutinated by anti-A, and B+ the event that they are agglutinated by anti-B. Thus, P(A+) = 0.44, P(B+) = 0.14. If these two events are independent, we should expect

$$P(A+\ B+) = P(A+)\,P(B+) = (0.44)(0.14) = 0.06$$

In reality, less than 4% of Caucasians fall into this category (i.e., have an AB blood type); therefore, the two traits are not distributed independently in the population. This finding was the first line of evidence used to argue that these two traits are due to multiple alleles at one locus (the ABO locus), rather than to segregation at two separate loci.

NONINDEPENDENCE DUE TO LINKAGE DISEQUILIBRIUM

The white cells of 31% of the Caucasian population react positively with HLA anti-A1, and those of 21% react positively with HLA anti-B8 (HLA denotes the human leukocyte antigen system). If these two traits are independent, we would expect the proportion of the population whose cells would react positively to both antisera to be $(0.31)(0.21) = 0.065$ (i.e., 6.5%). In fact, we find that 17% of the population are both A1 positive and B8 positive. In this case family studies have shown that two loci (A and B), very close together on chromosome 6, are involved. Nonindependence between two loci that are close together is termed *linkage disequilibrium*.

These two examples illustrate the fact that more than one biological phenomenon can lead to a lack of independence. In many cases in the literature, nonindependence is established, and then, on the basis of that evidence alone, a particular biological mechanism is incorrectly inferred.

CONDITIONAL PROBABILITY IN GENETIC COUNSELING

We shall now consider a very simple genetic example that will help you learn how to manipulate conditional probabilities. First, recall the formula for the joint probabilities of events A and B, which can be written (backward) as

$$P(A)\,P(B \mid A) = P(A \text{ and } B)$$

Now divide both sides by P(A) [which may be done provided P(A) is not zero], and we find

$$P(B \mid A) = \frac{P(A \text{ and } B)}{P(A)}$$

This gives us a formula for calculating the conditional probability of B given A, $P(B \mid A)$, if we know P(A and B) and P(A). Similarly, if we know P(A and B) and P(B), we can find

$$P(A|B) = \frac{P(A \text{ and } B)}{P(B)} \quad \text{[assuming P(B) is not zero]}$$

Now consider, as an example, hemophilia—a rare X-linked recessive disease in which there is a defect in the blood-clotting system. A carrier mother who marries an unaffected man transmits the disease to half her sons and to none of her daughters (half her daughters will be carriers, but none will be affected with the disease). What is the probability that she will bear an affected child? The child must be either a son or a daughter, and these are mutually exclusive events. Therefore,

$$\begin{aligned} P(\text{affected child}) &= P(\text{affected son}) + P(\text{affected daughter}) \\ &= P(\text{son and affected}) + P(\text{daughter and affected}) \\ &= P(\text{son})P(\text{affected}|\text{son}) + \\ &\quad\; P(\text{daughter})P(\text{affected}|\text{daughter}) \end{aligned}$$

Assume a 1:1 sex ratio [i.e., P(son) = P(daughter) = ½], and use the fact that P(affected|son) = ½. Furthermore, in the absence of mutation (which has such a small probability that we shall ignore it), P(affected|daughter) = 0. We therefore have

$$P\ (\text{affected child}) = \frac{1}{2} \times \frac{1}{2} + \frac{1}{2} \times 0 = \frac{1}{4}$$

Now suppose an amniocentesis is performed and thus the gender of the fetus is determined. If the child is a daughter, the probability of her being affected is virtually zero. If, on the other hand, the child is a son, the probability of his being affected is one half. Note that we can derive this probability by using the formula for conditional probability:

$$P\ (\text{affected}|\text{son}) = \frac{P(\text{son and affected})}{P(\text{son})} = \frac{\frac{1}{4}}{\frac{1}{2}} = \frac{1}{2}$$

Of course in this case you do not need to use the formula to obtain the correct answer, but you should nevertheless be sure to understand the details of this example. Although a very simple example, it illustrates how one proceeds in more complicated examples. Notice that knowledge of the sex of the child changes the probability of having an affected child from ¼ (before the gender was known) either to ½ (in the case of a son) or to 0 (in the case of a daughter). Conditional probabilities are used extensively in genetic counseling. We shall now discuss the use of a very general theorem that gives us a mechanism for adjusting the probabilities of events as more information becomes available.

BAYES' THEOREM

The Englishman Thomas Bayes wrote an essay on probability that he was hesitant to publish because he recognized the flaw in assuming, as he did in his essay, that all possible outcomes are equally likely. The essay was nevertheless published in 1763, after his death, by a friend. What is now called **Bayes' theorem** evolved from this essay and does not contain the original flaw. The theorem gives us a method of calculating new probabilities to take account of new information. Suppose that 20% of a particular population has a certain disease, D. For example, the disease might be hypertension, defined as having an average diastolic blood pressure of 95 mm Hg or greater taken once a day over a period of 5 days. In Figure 4–2 we represent the whole population by a square whose sides are unity. The probability that a person has the disease, P(D), and the probability that a person does not have the disease, $P(\overline{D}) = 1 - P(D)$, are indicated along

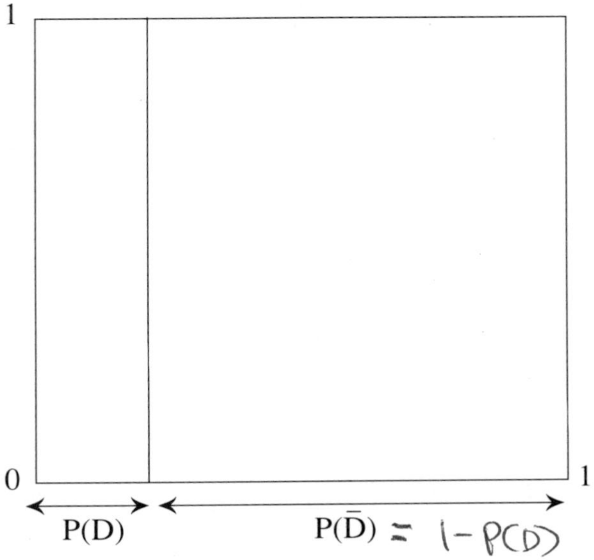

Figure 4–2. In the whole population, represented by a square whose sides are unity, the probability of having the disease is P(D), and of not having the disease is $P(\overline{D})$, as indicated along the horizontal axis.

the bottom axis. Thus the areas of the two rectangles are the same as these two probabilities.

Now suppose we have a test that picks up a particular symptom S associated with the disease. In our example, the test might be to take just one reading of the diastolic blood pressure, and S might be defined as this one pressure being 95 mm Hg or greater. Alternatively, we could say that the test result is positive if this one blood pressure is 95 mm Hg or greater, negative otherwise. Before being tested, a random person from the population has a 20% probability of having the disease. How does this probability change if it becomes known that the symptom is present?

Assume that the symptom is present in 90% of all those with the disease but only 10% of all those without the disease, i.e., $P(S|D) = 0.9$ and $P(S|\overline{D}) = 0.1$. In other words, the sensitivity and the specificity of the test are both 0.9. These conditional probabilities are indicated along the vertical axis in Figure 4–3. The gray rectangles represent the joint probabilities that the symptom is present and that the disease is present or not:

$$P(S \text{ and } D) = P(D)\,P(S|D) = (0.2)(0.9) = 0.18$$
$$P(S \text{ and } \overline{D}) = P(\overline{D})\,P(S|\overline{D}) = (0.8)(0.1) = 0.08$$

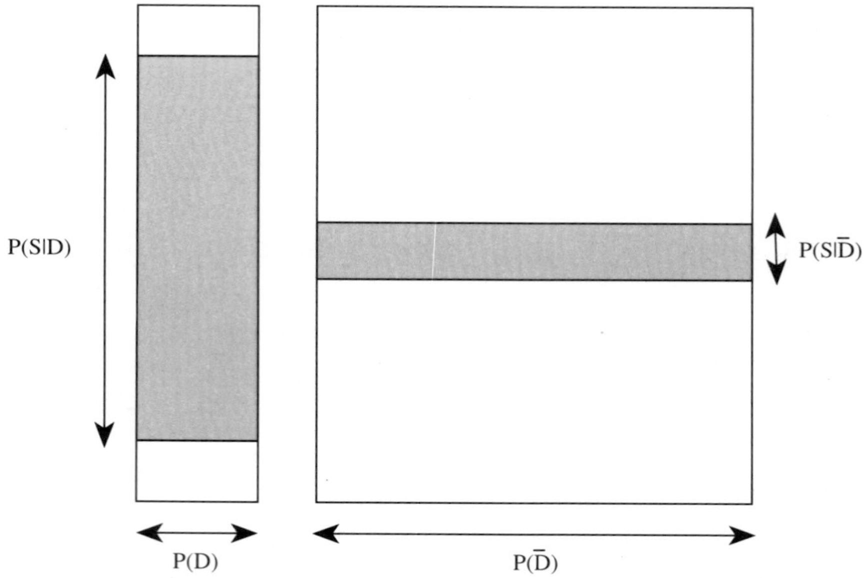

Figure 4–3. Within each subpopulation, D and $\overline{D}$, the conditional probability of having a positive test result or symptom, S, is indicated along the vertical axis. The gray rectangles represent the joint probabilities $P(D)P(S|D) = P(S,D)$ and $P(\overline{D})P(S|\overline{D}) = P(S,\overline{D})$.

If we know that the symptom is present, then we know that only the gray areas are relevant, i.e., we can write, symbolically,

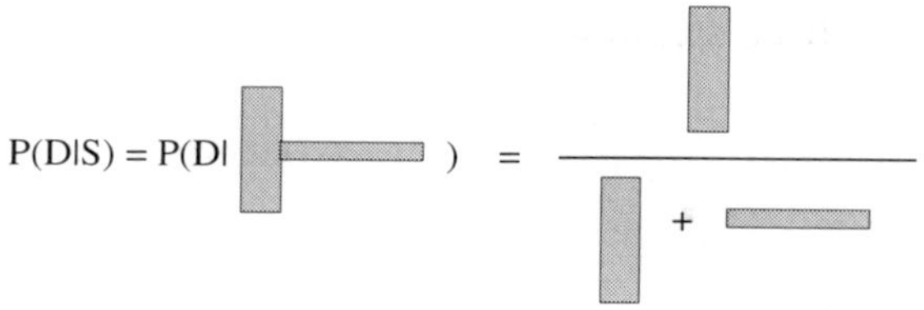

$$= \frac{P(S \text{ and } D)}{P(S \text{ and } D) + P(S \text{ and } \bar{D})} = \frac{P(D)P(S|D)}{P(D)P(S|D) + P(\bar{D})P(S|\bar{D})}$$

$$= \frac{(0.2)(0.9)}{(0.2)(0.9) + (0.8)(0.1)} = 0.69\,,$$

which is the positive predictive value of the test. This, in essence, is Bayes' theorem. We start with a *prior* probability of the disease, P(D), which is then converted into a *posterior* probability, $P(D|S)$, given the new knowledge that the symptom S is present.

More generally, we can give the theorem as follows. Let the new information that is available be that the event S occurred. Now suppose the event S can occur in any one of k distinct, mutually exclusive ways. Call these ways $D_1, D_2, \ldots, D_k$ (in the above example there were just two ways, the person either had the disease or did not have the disease; in general there may be k alternative diagnoses possible). Suppose that with no knowledge about S, these have prior probabilities $P(D_1)$, $P(D_2)$, . . ., and $P(D_k)$, respectively. Then the theorem states that the *posterior* probability of a particular D, say D_j, conditional on S having occurred, is

$$P(D_j|S) = \frac{P(D_j)P(S|D_j)}{P(D_1)P(S|D_1) + P(D_2)P(S|D_2) + \cdots + P(D_k)P(S|D_k)}$$
$$= \frac{P(D_j \text{ and } S)}{P(D_1 \text{ and } S) + P(D_2 \text{ and } S) + \cdots + P(D_k \text{ and } S)}$$

The theorem can thus be remembered as "the joint probability divided by the sum of the joint probabilities" (i.e., the posterior probability of a par-

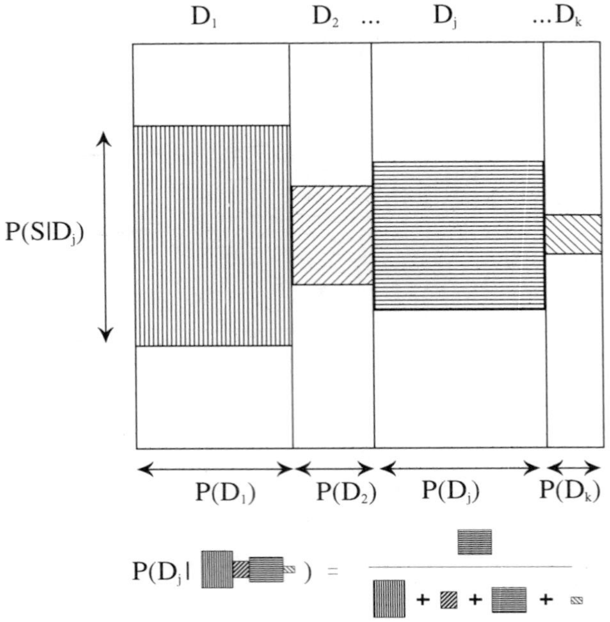

Figure 4–4. Bayes' theorem. The probabilities of various diagnoses, D_1, D_2,. . . D_j,. . . D_k, are indicated on the horizontal axis and the conditional probability of a particular symptom, within each diagnostic class D_i, is indicated on the vertical axis. Thus each hatched rectangle is the joint probability of the symptom and a diagnostic class.

ticular D, given that S has occurred, is equal to the joint probability of D and S occurring, divided by the sum of the joint probabilities of each of the Ds and S occurring). This is illustrated in Figure 4–4.

In order to illustrate how widely applicable Bayes' theorem is, we shall consider some other examples. First, suppose that a woman knows that her mother carries the gene for hemophilia (because her brother and her maternal grandfather both have the disease). She is pregnant with a male fetus and wants to know the probability that he will be born with hemophilia. There is a ½ probability that she has inherited the hemophilia gene from her mother, and a ½ probability that (given that she inherited the gene) she passes it on to her son. Thus, if this is the only information available, the probability that her son will have hemophilia is ¼. This is illustrated in Figure 4–5*a*. Now suppose we are told that she already has a son who does not have hemophilia (Fig. 4–5*b*). What is now the probability that her next son will be born with hemophilia? Is it the same, is it greater than ¼, or is it less than ¼? That she has already had a son without hemophilia is new information that has a direct bearing on the situation. In order to see this, consider a more extreme situation: suppose she has already had 10 unaffected sons. This would suggest that she did not inherit the gene

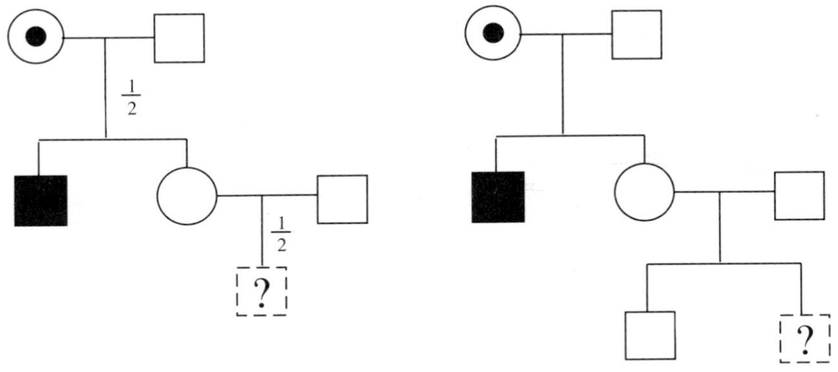

Figure 4–5. Woman (○) with a hemophilic brother (■) and a carrier mother (⊙), married to a normal male (□), is pregnant with a male fetus (?), for whom we want to calculate the probability of having hemophilia: (a) with no other information, (b) when she already has a son who does not have hemophilia.

from her mother, and if that is the case she could not pass it on to her future sons. If, on the other hand, she already has had a son with hemophilia, she would know without a doubt that she had inherited the gene from her mother and every future son would have a ½ probability of being affected. Thus, the fact that she has had one son who is unaffected *decreases* the probability that she inherited the hemophilia gene, and hence the probability that her second son will be affected. We shall now use Bayes' theorem to calculate the probability that the woman inherited the gene for hemophilia, given that she has a son without hemophilia. We have $k = 2$ and define the following events:

S = the woman has an unaffected son
D_1 = the woman inherited the hemophilia gene
D_2 = the woman did not inherit the hemophilia gene

Before we know that she has an unaffected son, we have the prior probabilities

$$P(D_1) = P(D_2) = \frac{1}{2}$$

Given whether or not she inherited the hemophilia gene, we have the conditional probabilities

$$P(S|D_1) = \frac{1}{2} \qquad P(S|D_2) = 1$$

Therefore, applying Bayes' theorem, the posterior probability that she inherited the hemophilia gene is

$$P(D_1|S) = \frac{P(D_1)P(S|D_1)}{P(D_1)P(S|D_1) + P(D_2)P(S|D_2)}$$

$$= \frac{\frac{1}{2}\cdot\frac{1}{2}}{\frac{1}{2}\cdot\frac{1}{2}+\frac{1}{2}\cdot 1} = \frac{1}{3}$$

Thus, given all that we know, the probability that the woman inherited the hemophilia gene is ⅓. Therefore, the probability that her second son is affected, given all that we know, is one half of this, i.e., ⅙. Note that P(S) = ⅙ is less than the ¼ probability that would have been appropriate if she had not already had a normal son. In general, the more unaffected sons we know she has, the less the probability that her next son will be affected.

Let us take as another example a situation that could arise in paternity testing. Suppose a mother has blood type A (and hence genotype AA or AO), and her child has blood type AB (and hence genotype AB). Thus, the child's A gene came from the mother and the B gene must have come from the father. The mother alleges that a certain man is the father, and his blood is typed. Consider two possible cases:

CASE 1

The alleged father has blood type O, and hence genotype OO. Barring a mutation, he could not have been the father: this is called an exclusion. Where there is an exclusion, we do not need any further analysis.

CASE 2

The alleged father has blood type AB, and therefore genotype AB, which is relatively rare in the population. Thus, not only *could* the man be the father, but he is also *more likely* to be the father than a man picked at random (who is much more likely to be O or A, and hence not be the father). In this case we might wish to determine, on the basis of the evidence available, the probability that the alleged father is in fact the true father. We can use Bayes' theorem for this purpose, provided we are prepared to make certain assumptions:

1. Assume there have been no blood-typing errors and no mutation, and that we have the correct mother. It follows from this assumption that the following two events occurred: (1) the alleged father is AB; and (2) the child received B from the true father—call this event S.

2. Assume we can specify the different, mutually exclusive ways in which S could have occurred. For the purposes of this example we shall suppose there are just two possible ways in which event S could have occurred (i.e., $k = 2$, again in the theorem):

(1) The alleged father is the true father (D_1). We know from mendelian genetics that the probability of a child receiving a B gene from an AB father is 0.5, i.e., $P(S|D_1) = 0.5$.

(2) Or a random man from a specified population is the true father (D_2). We shall assume that the frequency of the gene B in this population is 0.06, so that $P(S|D_2) = 0.06$. The probability that a B gene is passed on to a child by a random man from the population is the same as the probability that a random gene in the population is B (i.e., the population gene frequency of B).

3. Assume we know the prior probabilities of these two possibilities. As explained in Appendix 1, these could be estimated from the previous experience of the laboratory doing the blood typing. We shall assume it is known that 65% of alleged fathers whose cases are typed in this laboratory are in fact the true fathers of the children in question, i.e., $P(D_1) = 0.65$ and $P(D_2) = 0.35$.

We are now ready to use the formula substituting $P(D_1) = 0.65$, $P(D_2) = 0.35$, $P(S|D_1) = 0.5$ and $P(S|D_2) = 0.06$. Thus we obtain

$$P(D_1|S) = \frac{(0.65)(0.5)}{(0.65)(0.5) + (0.35)(0.06)}$$
$$= \frac{0.325}{0.325 + 0.021}$$
$$= 0.94$$

A summary of this application of Bayes' theorem is shown by means of a tree diagram in Figure 4–6.

Thus, whereas the alleged father, considered as a random case coming to this particular paternity testing laboratory, originally had a 65% probability of being the true father, now, when we take the result of this blood typing into consideration as well, he has a 94% probability of being the true father. In practice, many different blood systems are determined, which usually result in either one or more exclusions or a very high probability of paternity. We can never prove that the alleged father is the true father, because a random man could also have the same blood type as the true father. But if enough genetic systems are typed, either an exclusion will be found or the final probability will be very close to

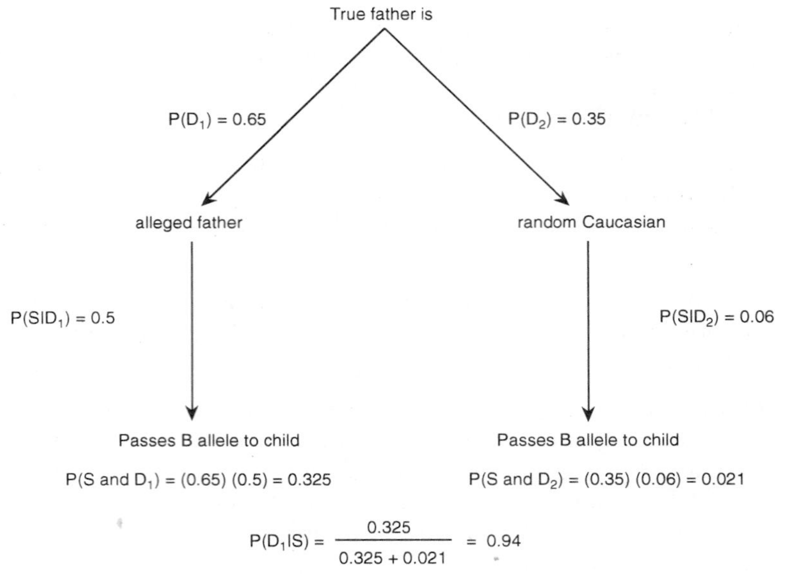

Figure 4–6. Calculation of the probability that the father of a child is the alleged father, on the basis of ABO blood types, using Bayes' theorem.

unity. In fact it is possible, using the many new genetic systems that have been discovered in DNA (obtainable from the white cells of a blood sample), to exclude virtually everyone except the monozygotic twin of the true father.

Several points should be noted about the use of Bayes' theorem for calculating the "probability of paternity." First, when the method was initially proposed, it was assumed that $P(D_1) = P(D_2) = 0.5$ (and in fact, there may be many who still misguidedly make this assumption). The argument given was that the alleged father was either the true father or was not, and in the absence of any knowledge about which was the case, it would seem reasonable to give these two possibilities equal prior probabilities. To see how unreasonable such an assumption is, consider the following. If I roll a die, it will come up either a "six" or "not a six." Since I do not know which will happen, should I assume equal probabilities for each of these two possibilities? In fact, of course, previous experience and our understanding of physical laws suggest that the probability that a fair die will come up a "six" is 1/6, whereas the probability it will come up "not a six" is 5/6. Similarly we should use previous experience and our knowledge of genetic theory to come up with a reasonable prior probability that the alleged father is the true father. We can understand now why Thomas Bayes was very unhappy with the idea that lack of knowledge about prior probabilities implies that they are equal.

Second, although the probability of paternity obtained in this way may be perfectly valid, it may not be relevant for the particular man in question. A blood-typing laboratory may validly calculate a 99% probability of paternity for 100 different men, and exactly one of these men may not be the true father of the 100 children concerned. But if it were known, for that one man, that he could not have had access to the woman in question, the relevant prior probability for him would be 0. It was a step forward when blood-typing evidence became admissible in courts of law, but it would be a step backward if, on account of this, other evidence were ignored. The so-called probability of paternity summarizes the evidence from the blood-typing only.

Last, always remember that probabilities depend on certain assumptions. At the most basic level, a probability depends on the population of reference. Thus when we assumed $P(S|D_2) = 0.06$, we were implicitly assuming a population of men in which the relative frequency of the B gene is 0.06—which is appropriate for Caucasians but not necessarily for other racial groups. Had we not wished to make this assumption, but rather that the father could have been from one of two different racial groups, it would have been necessary to assume the specific probabilities that he came from each of those racial groups. A posterior probability obtained using Bayes' theorem also assumes we know all the possible ways that the event B can occur and can specify an appropriate prior probability for each. In our example, we assumed that the true father was either the alleged father—the accused man—or a random Caucasian man. But could the true father have been a relative of the woman? Or a relative of the accused man—perhaps his brother? We can allow for these possibilities when we use Bayes' theorem, because in the general theorem we are not limited to $k = 2$, but then we must have an appropriate prior probability for each possibility. In practice, it may be difficult to know what the appropriate prior probabilities are, even if we are sure that no relatives are involved.

The examples given above have been kept simple for instructive purposes. Nevertheless, you should begin to have an idea of the powerful tool provided by Bayes' theorem. It allows us to synthesize our knowledge about an event to update its probability as new knowledge becomes available. With the speed of modern computers it is practical to perform the otherwise tedious calculations even in a small office setting.

LIKELIHOOD RATIO

If, in paternity testing, we do not know and are unwilling to assume particular prior probabilities, it is impossible to derive a posterior probability. But we could measure the strength of the evidence that the alleged father is the true father by the ratio $P(S|D_1)/P(S|D_2) = 0.5/0.06 = 8.3$, which in this particular situation is called the "paternity index." It is simply the probability of what we observed (the child receiving allele B from the true father) if the alleged father is in fact the father, *relative to* the probability of what we observed if a random Caucasian man is the father. This is an example of what is known as a **likelihood ratio.** If we have any two hypotheses D_1 and D_2 that could explain a particular event S, then the likelihood ratio of D_1 relative to D_2 is defined as the ratio $P(S|D_1)/P(S|D_2)$. The conditional probability $P(S|D_1)$—the probability of observing S if the hypothesis D_1 ("the alleged father is the true father") is true—is also called the "likelihood" of the hypothesis D_1 on the basis of S having occurred. Similarly, $P(S|D_2)$ would be called the likelihood of D_2. The likelihood ratio is a ratio of two conditional probabilities and is used to assess the relative merits of two "conditions" (D_1 versus D_2), or hypotheses. This concept has many applications in statistics and we shall discuss it further in Chapter 7.

SUMMARY

1. The probability of event A is classically defined as the number of outcomes that are favorable to A divided by the total number of possible outcomes. This definition has the disadvantage that it requires one to assume that all possible outcomes are equiprobable. The frequency definition assumes one can perform a trial many times and defines the probability of A as the limiting proportion, as the number of trials tends to infinity, of the trials in which A occurs. The axiomatic defi-

nition of probability is simply a set of positive numbers that sum to unity.

2. A valid probability need not be the clinically relevant probability; the patient at hand may belong to a special subset of the total population to which the probability refers.

3. The probability of A or B occurring is given by

$$P(\text{A or B}) = P(\text{A}) + P(\text{B}) - P(\text{A and B})$$

If A and B are mutually exclusive events, $P(\text{A and B}) = 0$.

4. The joint probability of A and B occurring is given by

$$P(\text{A and B}) = P(\text{A})P(\text{B}|\text{A}) = P(\text{B})P(\text{A}|\text{B})$$

If A and B are independent, $P(\text{A and B}) = P(\text{A})P(\text{B})$, and conversely, if $P(\text{A and B}) = P(\text{A})P(\text{B})$, then A and B are independent. Many different biological mechanisms can be the cause of dependence. Mutually exclusive events are never independent.

5. The conditional probability of A given B is given by

$$P(\text{A}|\text{B}) = \frac{P(\text{A and B})}{P(\text{B})}$$

6. Bayes' theorem states that the posterior probability that a particular D_j occurred, after it is known that the event S has occurred, is equal to the joint probability of D_j and S divided by the sum of the joint probabilities of each possible D and S:

$$P(D_j|S) = \frac{P(D_j \text{ and } S)}{P(D_1 \text{ and } S) + P(D_2 \text{ and } S) + \cdots + P(D_k \text{ and } S)}$$

$$= \frac{P(D_j)P(S|D_j)}{P(D_1)P(S|D_1) + P(D_2)P(S|D_2) + \cdots + P(D_k)P(S|D_k)}$$

It is assumed that we can specify a complete set of mutually exclusive ways in which S can occur, together with the prior probability of each. It does not assume that these prior probabilities are all equal.

7. When the prior probabilities of D_1 and D_2 are unknown, we can consider $P(S|D_1)/P(S|D_2)$, the likelihood ratio of D_1 versus D_2, as a summary of the evidence provided by the event S relative to D_1 and D_2.

FURTHER READING

Inglefinger, J.A., Mosteller, F., Thibodeaux, L.A., and Ware, J.H. (1983) Biostatistics in Clinical Medicine. Macmillan, New York. (Chapter 1 gives several good examples of probability applied to clinical cases.)

Weinstein, M.C., Fineberg, H.V., Elstein, A.S., Frazier, H.S., Neuhauser, D., Neutra, R.R., and McNeil, B.J. (1980) Clinical Decision Analysis. W.B. Saunders, Philadelphia. (Section 4.9 gives an interesting illustration of Bayes' theorem applied to emergency-ward patients with acute abdominal pain. It is seen how the probability that such a patient has appendicitis, pancreatitis, or nonspecific abdominal disease changes after the result of a simple clinical test, rebound tenderness, is known.)

Mendenhall, W., Wackerly, D.D., and Scheaffer, R.L. (1990) Mathematical Statistics with Applications, 4th ed. PWS-Kent Publishing Company, Boston. (Although written at a more mathematical level, the first few chapters contain many examples and exercises on probability.)

PROBLEMS

Problems 1 through 4 are based on the following: For a particular population, the lifetime probability of contracting glaucoma is approximately 0.007 and the lifetime probability of contracting diabetes is approximately 0.020. A researcher finds (for the same population) that the probability of contracting both of these diseases in a lifetime is 0.0008. A and B

1. What is the lifetime probability of contracting either glaucoma or diabetes?
2. What is the lifetime probability of contracting glaucoma for a person who has, or will have, diabetes?
3. What is the lifetime probability of contracting diabetes for a person who has, or will have, glaucoma?

   Possible answers for problems 1 through 3 are

   A. 0.0400
   B. 0.0278
   C. 0.0296
   D. 0.0262
   E. 0.1143

4. On the basis of the information given, which of the following conclusions is most appropriate for the two events: contracting glaucoma and contracting diabetes? They

   A. are independent
   B. are not independent

C. have additive probabilities
D. have genetic linkage
E. have biological variability

5. A certain operation has a fatality rate of 30%. If this operation is performed independently on three different patients, what is the probability that all three operations will be fatal?

A. 0.09
B. 0.90
C. 0.009
D. 0.027
E. 0.27

6. The probability that a certain event A occurs in a given run of an experiment is 0.3. The outcome of each run of this experiment is independent of the outcomes of other runs. If the experiment is run repeatedly until A occurs, what is the probability exactly four runs will be required?

A. 0.0531
B. 0.1029
C. 0.2174
D. 0.4326
E. 0.8793

$P(A) = .3$

.0008

$P(A) \ f$

7. A small clinic has three physicians on duty during a standard work week. The probabilities that they are absent from the clinic at any time during a regular work day because of a hospital call are, respectively, 0.2, 0.1 and 0.3. If their absences are independent events, what is the probability that at least one physician will be in the clinic at all times during a regular workday? (Disregard other types of absences.)

A. 0.006
B. 0.251
C. 0.496
D. 0.813
E. 0.994

$1 - (.2 \times .1 \times .3) =$

$1 -.$

8. If two thirds of patients survive their first myocardial infarction and one third of these survivors is still alive 10 years after the first attack, then among all patients who have a myocardial infarction, what proportion will die within 10 years of the first attack? (Hint: Draw a tree diagram.)

A. ⅑
B. ⅔
C. ⅓

$\frac{x}{3}$ died $\boxed{2/3}$ survived
10 years

$\frac{1}{3}$ died   ⅓·alive

D. ⅔
E. 7/9

9. If 30% of all patients who have a certain disease die during the first year and 20% of the first-year survivors die before the fifth year, what is the probability an affected person survives past 5 years? (*Hint:* Draw a tree diagram.)
   A. 0.50
   B. 0.10
   C. 0.56
   D. 0.06
   E. 0.14

10. Suppose that 5 men out of 100 and 25 women out of 10,000 are colorblind. A colorblind person is chosen at random. What is the probability the randomly chosen person is male? (Assume males and females to be in equal numbers.)
   A. 0.05
   B. 0.25
   C. 0.75
   D. 0.95
   E. 0.99

11. A mother with blood type B has a child with blood type O. She alleges that a man whose blood type is O is the father of the child. What is this likelihood that the man is the true father, based on this information alone, relative to a man chosen at random from a population in which the frequency of the O gene is 0.67?
   A. 0.33
   B. 1.49
   C. 2.00
   D. 0.50
   E. 0.67

12. There is a 5% chance that the mother of a child with Down syndrome has a particular chromosomal translocation, and if she has that translocation, there is a 16% chance that a subsequent child will have Down syndrome; otherwise the chance of a subsequent child having Down syndrome is only 1%. Given these facts, what is the probability, for a woman with a Down syndrome child, that her next child has Down syndrome?
   A. 0.21
   B. 0.16
   C. 0.05
   D. 0.02
   E. 0.01

13. A person randomly selected from a population of interest has a probability of 0.01 of having a certain disease which we shall denote D. The probability of a symptom S, which may require a diagnostic test to evaluate its presence or absence, is 0.70 in a person known to have the disease. The probability of S in a person known not to have the disease is 0.02. A patient from this population is found to have the symptom. What is the probability this patient has the disease?

A. 0.01
B. 0.02
C. 0.26
D. 0.53
E. 0.95

14. When the prior probabilities of D_1 and D_2 are unknown, the quantity $P(S|D_1)/P(S|D_2)$ is called

A. the risk of S given D_2
B. the correlation ratio attributable to D_1 and D_2
C. Bayes' theorem
D. the joint probability of D_1 and D_2 occurring in the presence of S
E. the likelihood ratio of the evidence provided by S for D_1 relative to D_2

15. Let E be the event ''exposed to a particular carcinogen,'' N the event ''not exposed to the carcinogen,'' and D the event ''disease present.'' If the likelihood ratio $P(D|E)/P(D|N)$ is 151.6, this can be considered to be a summary of the evidence that

A. disease is more likely to occur in the exposed
B. exposure is more probable in the diseased
C. the conditional probability of disease is less than the unconditional probability
D. the conditional probability of disease is equal to the unconditional probability
E. the events D and E are mutually exclusive

CHAPTER FIVE

Key Concepts

variable, random variable, response variable, variate
parameter
probability (density) function, cumulative probability distribution function
binomial, Poisson, uniform, and normal (gaussian) distributions
standardized, standard normal distribution

Symbols and Abbreviations

e	irrational mathematical constant equal to about 2.71828
f(y)	(probability) density function
F(y)	cumulative (probability) distribution function
n	sample size; parameter of the binomial distribution
p	sample proportion
x, y	particular values of (random) variables
X, Y	random variables
Z	a standardized normal, random variable
λ	parameter of the Poisson distribution (Greek letter lambda)
π	population proportion (Greek letter pi)
μ	population mean (Greek letter mu)
σ	population standard deviation (Greek letter sigma)
!	factorial

Random Variables and Distributions

VARIABILITY AND RANDOM VARIABLES

A general term for any characteristic or trait we might measure on a patient or other unit of study is **variable.** A patient's height, weight, and blood pressure are examples of variables. These traits are variable (and hence called variables) for two main reasons:

1. Measurement error causes the values we measure to differ, even though we may be repeatedly measuring exactly the same thing. Broadly speaking, measurement error includes errors of misclassification, variability in the measuring instruments we use, and variability among observers in how they read those instruments. For example, no two mercury sphygmomanometers are exactly the same and their readings may be slightly different. Even if a group of doctors all view the same film of the movement of mercury in such an instrument, together with listening to the sounds heard in the stethoscope, they will nevertheless not all agree on how they read the diastolic and systolic blood pressures. Similarly, if identical blood samples are sent for lipid analyses to two different laboratories—or even to the same laboratory but under different names—the results that are returned will be different.

2. There is inherent variability in all biological systems. There are differences among species, among individuals within a species, and among parts of an individual within an individual. No two human beings have exactly the same height and weight at all stages of their growth; no two

muscle cells have exactly the same chemical composition. The same individual, or the same muscle cell, changes with age, time of day, season of the year, and so forth. An individual's blood pressure, for example, can fluctuate widely from moment to moment depending on how fearful the subject is.

When a variable is observed as part of an experiment or a well-defined sampling process, it is usually called a **response variable, random variable,** or **variate.** Thus, we think of each observation in a set of data as the outcome of a random variable. For example, consider a family in which there is a given probability that a child is affected with some disease. Then whether or not a child is affected is the outcome of a discrete random variable. We often assign arbitrary numbers to the outcomes of such a random variable, for instance, 0 if not affected, 1 if affected. In other settings, variables such as blood pressure measurements and cholesterol levels are continuous random variables.

Data of all types, but especially those that involve observations on a large number of subjects, are difficult to interpret unless they are organized in a way that lends itself to our seeing general patterns and tendencies. A first and important way of organizing data to obtain an overview of the patterns they display is to construct a frequency distribution, as we described in Chapter 3. Here we shall be concerned with **probability distributions.** A probability distribution is a model for a random variable, describing the way the probability is distributed among the possible values the random variable can take on. As we saw in Chapter 4, probability can be interpreted as relative frequency in an indefinitely large number of trials. The distributions we shall now describe are thus theoretical ones, but nevertheless ones that are of great practical importance and utility. Mathematically, the concepts "probability distribution" and "random variable" are interrelated, in that each implies the existence of the other; a random variable must have a probability distribution and a probability distribution must be associated with a random variable. If we know the probability distribution of a random variable, we have at our disposal information that can be extremely useful in studying the patterns and tendencies of data associated with that random variable. Many mathematical models have been developed to describe the different shapes a probability distribution can have. One broad class of such models has been developed for discrete random variables; a second class is for continuous random variables. In the following sections we describe just a few common models to introduce you to the concept.

BINOMIAL DISTRIBUTION

= 2 parameters n & π

Let us start with a simple dichotomous trait (a qualitative trait that can take on one of only two values). Consider a rare autosomal dominant condition such as achondroplasia (a type of dwarfism). Let the random variable we are interested in be the number of children with achondroplasia in a family. Suppose a couple, one of whom has achondroplasia, has a single child. Then the random variable is dichotomous (it must be 0 or 1), and the probability the child is unaffected is ½. We thus have the probability distribution for a random variable that has values 0 and 1 with probabilities given by

$$P(0) = \frac{1}{2}$$

$$P(1) = \frac{1}{2}$$

which we can graph as in Figure 5–1(*a*). If we call the random variable Y, we can also write this as

$$P(Y = 0) = \frac{1}{2}$$

$$P(Y = 1) = \frac{1}{2}$$

This completely describes the probability distribution of the random variable Y for this situation. Note that the distribution is described by a function of Y. This function, which gives us a rule for calculating the probability of any particular value of Y, is called a **probability function.**

Now suppose the couple has two children; then the random variable Y, the number affected, can take on three different values: 0, 1, or 2. (Note that this is a discrete random variable measured on an ordinal scale.) Using the laws of probability and the fact that each conception is an independent event (we shall exclude the possibility of monozygotic twins), we have

$$\begin{aligned} P(0) &= P(\text{1st child unaffected and 2nd child unaffected}) \\ &= P(\text{1st child unaffected}) \times \\ &\quad P(\text{2nd child unaffected} \mid \text{1st child unaffected}) \\ &= \frac{1}{2} \times \frac{1}{2} \\ &= \frac{1}{4} \end{aligned}$$

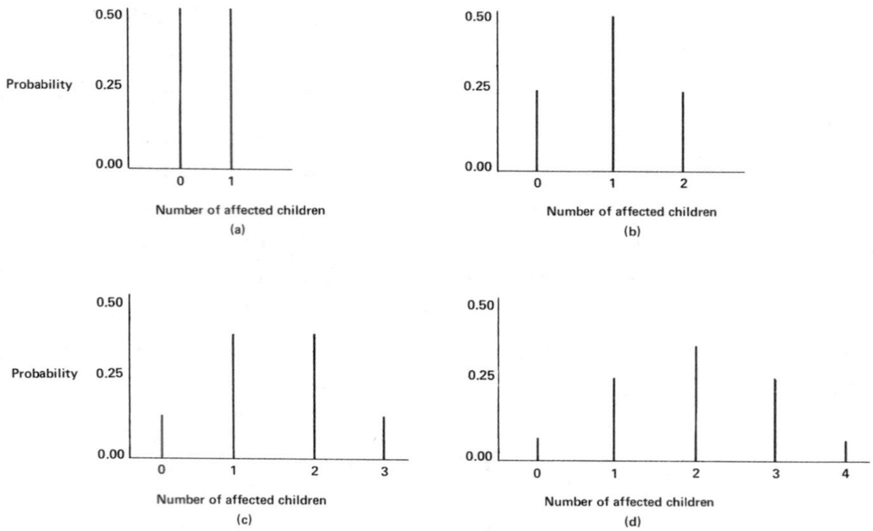

Figure 5–1. Probability distribution of number of children affected when the probability of each child being affected is ½ and the total number of children in a family is (a) 1, (b) 2, (c) 3, and (d) 4.

$$
\begin{aligned}
P(1) &= P(\text{one child affected}) \\
&= P(\text{1st child affected and 2nd child unaffected}) \\
&\quad + P(\text{1st child unaffected and 2nd child affected}) \\
&= P(\text{1st child affected}) \times P(\text{2nd child unaffected} \mid \text{1st child affected}) \\
&\quad + P(\text{1st child unaffected}) \times \\
&\quad P(\text{2nd child affected} \mid \text{1st child unaffected}) \\
&= \frac{1}{2} \times \frac{1}{2} + \frac{1}{2} \times \frac{1}{2} \\
&= \frac{1}{2}
\end{aligned}
$$

$$
\begin{aligned}
P(2) &= P(\text{both children affected}) \\
&= P(\text{1st child affected}) \times P(\text{2nd child affected} \mid \text{1st child affected}) \\
&= \frac{1}{2} \times \frac{1}{2} \\
&= \frac{1}{4}
\end{aligned}
$$

Note carefully that there are two mutually exclusive ways in which Y can take on the value 1: either the first child is affected or the second child is affected, and the other is not. This is why we simply added the two probabilities.

In summary, for a family of two children, we have the probability function

$$P(0) = \frac{1}{4} \qquad P(1) = \frac{1}{2} \qquad P(2) = \frac{1}{4}$$

which is graphed in Figure 5–1(b).

Using the same kind of argument, we find that the probability function of Y in a family with three children is

$$P(0) = \frac{1}{8} \qquad P(1) = \frac{3}{8} \qquad P(2) = \frac{3}{8} \qquad P(3) = \frac{1}{8}$$

and for a family of four children we find

$$P(0) = \frac{1}{16} \qquad P(1) = \frac{1}{4} \qquad P(2) = \frac{3}{8} \qquad P(3) = \frac{1}{4} \qquad P(4) = \frac{1}{16}$$

These are graphed in Figures 5–1(c) and (d).

All the probability functions we have so far considered are special cases of the **binomial probability distribution,** or simply the **binomial distribution.** The expression "binomial distribution" is used to name a family of many different distributions, a particular member of this family being determined by the values of two (non-random) variables called **parameters.** Suppose we perform n independent trials, and at each trial the probability of a particular event is π. Then n and π are the parameters of the distribution. In the above examples n is the number of children in the family and π is the probability that each should have achondroplasia. The probability that the event occurs a total of y times is given by the binomial probability distribution with parameters n and π, and is expressed mathematically by the probability function

$$P(y) = \frac{n!}{y!(n-y)!}\, \pi^y (1 - \pi)^{n-y} \qquad \text{for } y = 0, 1, 2, \ldots, n$$

In this formula, read $n!$ as "n factorial," or "factorial n"; it is equal to $n(n-1)(n-2)\ldots(2)(1)$. For example, $3! = (3)(2)(1) = 6$. To use the formula, you may need to use the fact that $0!$ is defined to be unity (i.e., $0! = 1$). Also, remember that any number raised to the power zero is unity [e.g., $(\frac{1}{2})^0 = 1$]. Consider, for example, finding the probability that, in a family of four children with one achondroplastic parent, no children will

n = # independent trials (ex– # children in family)

π = probability of a particular event

be affected. Here the parameters are $n = 4$ and $\pi = ½$, and we want to evaluate the probability function for $y = 0$. Thus

$$P(0) = \frac{4!}{0!(4-0)!}\left(\frac{1}{2}\right)^{0}\left(1-\frac{1}{2}\right)^{4-0}$$
$$= \frac{4!}{4!}(1)\left(\frac{1}{2}\right)^{4}$$
$$= \left(\frac{1}{2}\right)^{4} = \frac{1}{16}$$

Note that, in general, $P(0) = (1 - \pi)^n$, that is, the probability of no events of the type being considered occurring in a total of n trials is $(1 - \pi)^n$. Conversely, the probability of more than zero (i.e., at least one) such events occurring must be the complement of ("complement of" means "one minus") this, that is, $1 - (1 - \pi)^n$: the number of such events occurring must be either zero or more than zero. Finally, remember that the binomial probability distribution is relevant for a prespecified number n of *independent* trials in each of which there is the *same* probability π of a particular event occurring. Binomial probabilities are often relevant in genetic counseling because each conception can be considered an independent event with the same probability of resulting in a child having a given condition.

A NOTE ABOUT SYMBOLS

You may have noticed that we originally used the capital letter Y to denote a random variable, but when we gave the formula for the binomial distribution we used the small letter y. Throughout this book we shall use capital Roman letters for random variables and the corresponding small Roman letters to denote a particular value of that random variable. Thus P(y) is a shorthand way of writing $P(Y = y)$, the probability that the random variable Y takes on the particular value y (such as 0, 1, 2, etc.). Making this distinction now will make it easier to understand some of the concepts we discuss in later chapters. Just remember that a capital letter stands for a random variable, which can take on different values, whereas a lowercase letter stands for a particular one of these different values. We shall also, for the most part, use Greek letters for unknown population parameters. Thus we shall use p for the proportion of affected children we might observe in a sample of children, but π for the theoretical probability that a child should be affected. (Since it is not universal, however, this convention will not be followed in all the problems.) The Greek letter π will also be used later in this chapter with its usual mathematical meaning of the ratio of the circumference of a circle to its diameter. In the next section we shall need to use the theoretical quantity that mathematicians always denote by the Roman letter e. Like π, e cannot be expressed as a rational number; it is approximately 2.71828.

POISSON DISTRIBUTION

Another important family of discrete distributions is the **Poisson distribution,** named after the French mathematician S.D. Poisson (1791–1840). This can often be used as a model for random variables such as radioactive counts per unit of time, the number of calls arriving at a telephone switchboard per unit of time, or the number of bacterial colonies per Petri plate in a microbiology study. If y is a particular value of a random variable that follows a Poisson distribution, then the probability that this value will occur is given by the formula

$$P(y) = \frac{\lambda^y e^{-\lambda}}{y!} \quad \text{for } y = 0, 1, 2, \ldots$$

where the parameter λ is the mean or average value of the random variable.

Whereas the binomial is a two-parameter (n and π) family of distributions, the Poisson is a one-parameter (λ) family. The Poisson distribution can be derived as a limiting case of the binomial and is sometimes called the distribution of "rare events." Suppose we have a binomial distribution, but the number of trials n is indefinitely large, while the probability π of a particular event at each trial approaches zero; the resulting distribution, provided the average number of events expected in n trials is a finite quantity, is the Poisson. Thus the Poisson distribution also assumes independent events, each with the same probability of occurring, but in addition it assumes that the total number of such events *could be* (though with very small probability) indefinitely large.

As an example, suppose it is known that in a large hospital, two patients a month, on an average, give incorrect billing information. What is the probability of a month with no patient giving incorrect billing information in this hospital? Or that there should be one patient with incorrect billing information in a given month? Or two, or three? The Poisson distribution can be used to give good approximations to answer these questions. In this example, $\lambda = 2$. Thus, using the above formula, we can calculate

$$P(0) = \frac{2^0 e^{-2}}{0!} = 0.1353$$

$$P(1) = \frac{2^1 e^{-2}}{1!} = 0.2706$$

$$P(2) = \frac{2^2 e^{-2}}{2!} = 0.2706$$

$$P(3) = \frac{2^3 e^{-2}}{3!} = 0.1804$$

and so on. Note that for the Poisson distribution the probability of zero events is in general $e^{-\lambda}$, and the probability of at least one event is $1 - e^{-\lambda}$.

Poisson probabilities are good approximations to binomial probabilities when n is large. Suppose, for example, that in the same hospital exactly 500 patients are billed each month. Then it would be appropriate to use the binomial distribution with $n = 500$ and $\pi = 2/500$, rather than the Poisson distribution. We would then find, for no incorrect billings,

$$P(0) = \left(1 - \frac{1}{250}\right)^{500}$$

$$= \left(\frac{249}{250}\right)^{500} = 0.1348$$

so we see that the Poisson approximation (0.1353) is good to three decimal places in this instance. The Poisson distribution is appropriate when n is

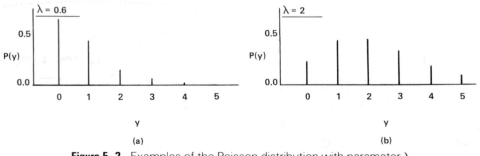

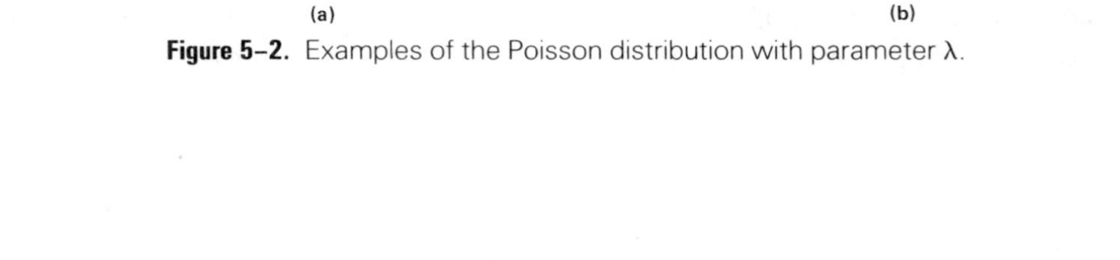

Figure 5–2. Examples of the Poisson distribution with parameter λ.

truly indefinitely large and so we have no idea what it is. This would be the case when we are measuring the number of radioactive counts in a unit of time, or when we are counting the number of red cells that fall in a square on a hemocytometer grid. Two examples of the Poisson distribution are shown in Figure 5–2. The first ($\lambda = 0.6$) was shown to give a good approximation to the distribution of the number of soldiers in the Prussian army killed by horse kicks in one year!

UNIFORM DISTRIBUTION

The two families of distributions we have considered so far—the binomial and the Poisson—are for discrete random variables. In each case we had a formula for P(y), the probability that the random variable Y takes on the value y, and we could plot these probabilities against y. A difficulty arises when we try to do this for a continuous random variable.

Consider as a model a spinner connected to a wheel that has equally spaced numbers ranging from 1 to 12 on its circumference, as on a clock (see Figure 5–3): If you spin the spinner (arrow), its tip is equally likely to stop at any point in the interval from 1 to 12 or, for that matter, at any point between any two consecutive numbers, say between 1 and 2. There are so many points on this circle that the probability the tip stops at any specific point is virtually zero. Yet, it is easy to see that the probability the tip stops between 1 and 2 is the same as the probability it stops between 9 and 10 (and, in fact, this probability is $\frac{1}{12}$). We can plot the probability distribution corresponding to this model as in Figure 5–4, in which the total area of the rectangle is unity: area = base × length = $12\left(\frac{1}{12}\right) = 1$. Moreover, the portion of the total area that lies between 1 and 2 is $\frac{1}{12}$, and this is the probability that the tip of the spinner stops between 1 and 2. Similarly, the area between 3 and 7 is $4\left(\frac{1}{12}\right) = \frac{1}{3}$, and this is the probability the spinner stops

Figure 5–3. Spinner wheel for a uniform distribution.

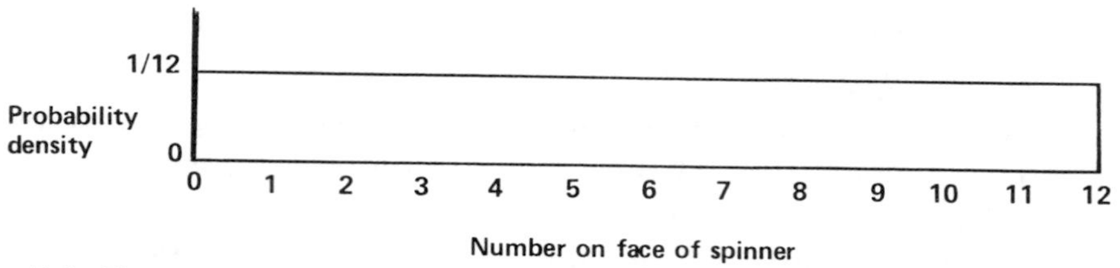

Figure 5–4. The uniform probability density corresponding to the spinner model in Figure 5–3.

between these two numbers. The height of the continuous line (1/12) is called the **probability density.**

The distribution described by the above model is called the **uniform** (or rectangular) distribution. It is one of the simplest for continuous data. We write the formula for the probability density as follows:

$$f(y) = \frac{1}{12}, \qquad 0 \leq y \leq 12$$

The random variable **Y** is the exact point between 0 and 12 at which the tip of the arrow comes to rest. The **probability density function** f(y) gives the height of the line plotted at each value y that **Y** can take on (in this case the height is always the same, 1/12, for all values of y between 0 and 12). We cannot call the height the probability of y because, as we have noted, the probability of any particular value of y—such as 2.5 or 7.68—is essentially zero. It is nevertheless a probability function, but we call it a probability density function, or simply a **density,** to stress that its values are not actual probabilities. The important thing to remember is that in the case of a continuous random variable, it is the *areas* under the probability density function f(y) that represent probabilities.

NORMAL DISTRIBUTION

The most important family of continuous densities used in statistical applications is the **normal** or **gaussian.** The latter name comes from the German astronomer K.F. Gauss (1777–1855). The density has a bell-shaped appearance and can also be shown to have a shape identical to that of a binomial distribution as n, the number of trials, becomes indefinitely large while π remains constant. Look again at the binomial distributions for $\pi = \frac{1}{2}$ and $n = 1, 2, 3$, and 4 in Figure 5–1, and then at the one in Figure

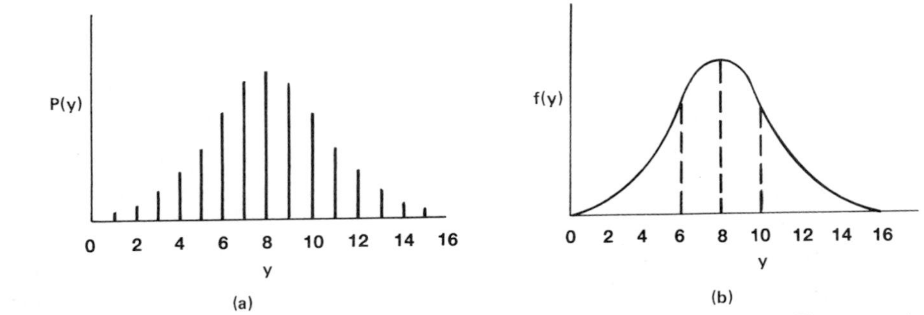

Figure 5–5. (a) The binomial probability function when $\pi = \frac{1}{2}$ and $n = 16$. (b) The normal density function when $\mu = 8$ and $\sigma = 2$.

5–5(a) for $\pi = \frac{1}{2}$ and $n = 16$. You will see that the shape is approaching the curve in Figure 5–5(b), which is a normal density.

The normal probability density function is symmetric (and hence has a coefficient of skewness equal to zero), with the highest point at the center. In Figure 5–5(b) the highest point occurs when y = 8, and this is the mean of the distribution—denoted by μ, the Greek letter mu. Although the normal density is always bell-shaped, it does not always have exactly the same amount of spread: it may be tall and thin, or it may be short and fat. It always, however, has two points of inflection, symmetrically placed about the mean. In Figure 5–5(b) these occur at y = 6 and y = 10; they are points at which the curve changes from being convex to concave, or from concave to convex. Starting at the extreme left, the curve rises more and more steeply until it reaches y = 6, at which point, although still rising, it does so more and more slowly. Similarly, at y = 10 the curve changes from decreasing more and more steeply to decreasing less and less steeply. The distance between the mean and the point of inflection (2 in Figure 5–5) measures the amount of spread of a normal density and is denoted by σ, the Greek letter sigma; it is the standard deviation of the distribution. The family of normal densities is thus a two-parameter family. The general formula for the curve is

$$f(y) = \frac{e^{-\frac{1}{2}\left(\frac{y-\mu}{\sigma}\right)^2}}{\sigma\sqrt{(2\pi)}} \qquad \text{for } -\infty \le y \le \infty$$

where π, e = usual mathematical constants
   y = value of random variable
   μ, σ = two parameters

[handwritten margin notes:] μ = mean

σ = standard dev. of distribution

Recall that ∞ is the mathematical symbol for "infinity"; thus the curve goes on "forever" in both directions. The properties that we have discussed, as well as those we shall discuss below, can all be deduced mathematically from this formula, which describes a family of normal distributions. If μ and σ are known, say $\mu = 8$ and $\sigma = 2$ as in Figure 5–5(*b*), then the formula describes just one member of the family. Other choices of μ and σ describe other members of the family. It is the value of μ that determines the *location* of the curve and the value of σ that determines the amount of *spread*. Nevertheless, all members of the family are bell-shaped and are described by the same general formula.

As with any other density, the total area under the normal curve is unity. Because it is symmetric, the probability to the left of the mean is ½, as is the probability to the right of the mean. The mean is thus the same as the median. The normal distribution has two other properties that you should commit to memory:

1. The probability that a normally distributed random variable lies within one standard deviation of the mean, that is, between $\mu - \sigma$ and $\mu + \sigma$ [between 6 and 10 in Figure 5–5(*b*)], is about two-thirds, or to be a little more precise, about 68%. It follows that there is a 16% probability it will lie below $\mu - \sigma$ and a 16% probability it will lie above $\mu + \sigma$ (16 + 68 + 16 = 100).

2. The probability that a normally distributed random variable lies within two standard deviations of the mean, that is, between $\mu - 2\sigma$ and $\mu + 2\sigma$ [between 4 and 12 in Figure 5–5(*b*)], is about 95%. It follows that there is about a 2½% probability it will lie below $\mu - 2\sigma$, and the same probability it will lie above $\mu + 2\sigma$ (2½ + 95 + 2½ = 100).

Theoretically, a normally distributed random variable can take on any value, positive or negative. Thus for the normal density pictured in Figure 5–5(*b*), for which $\mu = 8$ and $\sigma = 2$, it is theoretically possible (though highly improbable) that Y could be less than −1,000 or greater than 2,000. The actual probability, however, of being more than four or five standard deviations from the mean is virtually nil, and for all practical purposes can be ignored. For this reason we can often use the normal distribution to approximate real data, such as the heights of men, that we know cannot possibly be negative. Of course, there is nothing abnormal about a random variable not being normally distributed—remember that the word "normal" in this context means only that the density follows the bell-shaped curve that is mathematically defined above. But the normal distribution is of great practical importance in statistics because many types of data are approximately normally distributed or can be transformed to another scale (such as by taking logarithms or square roots) on which they are approximately normally distributed. Furthermore, even if the random variable of interest is not normally distributed and a normalizing transformation cannot readily be found, *sample averages of the random variable tend to become normally distributed as the sample size increases.* This important result holds true even if the random variable is of the categorical type. This result plays a fundamental role in many statistical analyses, because we often reduce our data to averages in an effort to summarize our findings. We shall come back to this important point later.

CUMULATIVE DISTRIBUTION FUNCTIONS

So far, we have described various distributions by probability functions we have denoted P(y) or f(y)—the former for discrete random variables (e.g., binomial, Poisson) and the latter for continuous random variables (e.g., uniform, normal). These same distributions can also be described by a different kind of function, namely the **cumulative probability distribution function,** or simply the **cumulative distribution.** Statisticians describe distributions both ways, and it is important to know which is being used. In this book we shall always include the word "cumulative," writing "cumulative probability distribution function" or simply "cumulative distribution" when that is meant. It will be understood that if the word "cumulative" is not stated, then a probability function is meant. (This is the usual convention in the articles you are likely to read, but not the convention in mathematical articles.)

For every probability function there is exactly one corresponding cumulative distribution function, and vice versa. Thus each gives us exactly the same information about the distribution of a random variable; they are two ways of expressing the same information. They are different mathematical functions, however, and they look quite different when we plot them. Whereas a probability function gives the probability or density of a particular value y of Y (it is analogous to a histogram or frequency polygon), the cumulative distribution function gives the probability that Y is less than or equal to a particular value y (it is analogous to a cumulative plot). The cumulative distribution function is particularly useful for continuous random variables, for which we cannot talk of the "probability of y."

The cumulative distribution functions for the uniform distribution illustrated in Figure 5–4 and the normal distribution illustrated in Figure 5–5(*b*) are depicted in Figure 5–6; they are denoted F(y).

First look at the cumulative uniform distribution. The probability that Y is less than or equal to 0 is 0, and we see F(0) = 0. The probability that Y lies between 0 and 1 (i.e., is less than or equal to 1) is $\frac{1}{12}$, and we see F(1) = $\frac{1}{12}$. The probability Y is less than or equal to 6 is ½, and we see F(6) = ½. Finally, Y must be less than or equal to 12, and in fact we see F(12) = 1. Although it is not plotted in the figure, clearly F(−1) = 0 and F(13) = 1. For any cumulative distribution it is always true that F(−∞) = 0 and F(∞) = 1.

Now look at the cumulative normal distribution illustrated in Figure 5–6(*b*). Note that it has the shape of a sloping, elongated "S," starting out close to zero and always below unity, except at infinity. Note that F(μ) = F(8) = 0.5, which corresponds to stating that the mean is equal to the median. Other cumulative probabilities can also be read off from this graph. Because there are an infinite number of possible graphs corresponding to the infinite number of members in the family of normal distributions, we would like a way to find probabilities for normal distributions without having to construct a cumulative distribution plot whenever the need arises. The next section provides a way of doing this.

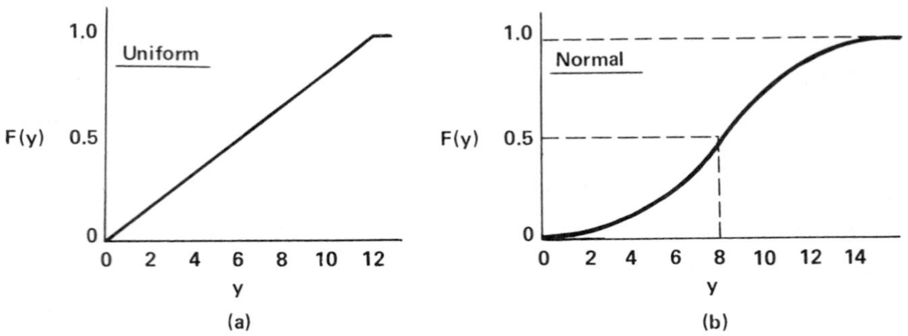

Figure 5–6. Cumulative distribution functions: (*a*) corresponding to the uniform density function in Figure 5–4; (*b*) corresponding to the normal density function in Figure 5–5.

THE STANDARD NORMAL (GAUSSIAN) DISTRIBUTION

Suppose a random variable Y is normally distributed in the population with mean μ and standard deviation σ. Let us subtract the mean from Y and divide the difference by the standard deviation. Call the result Z, i.e.,

$$Z = \frac{Y - \mu}{\sigma}$$

Then Z is also normally distributed, but has mean 0 and standard deviation 1. This is pictured in Figure 5–7. The distribution of Z is called the **standard normal distribution.** Any random variable that has mean 0 and standard deviation 1 is said to be **standardized.** If we know the mean and standard deviation of a random variable, we can always standardize that random variable so that it has a mean 0 and standard deviation 1. (The symbol Z, or the term "Z-score," is often used in the literature to indicate a variable that has been standardized—sometimes in an unspecified manner.) For a standardized *normal* random variable, about 68% of the population lies between −1 and +1, and about 95% of the population lies between −2 and +2. More accurate figures can be obtained from the cumulative standard normal distribution given in Table A2–1 (Appendix 2). For example, Table 1 indicates that F(−1) = 0.1587 and F(1) = 0.8413, so that the probability that a standard normal variable lies between −1 and 1 is

$$F(1) - F(-1) = 0.8413 - 0.1587 = 0.6826$$

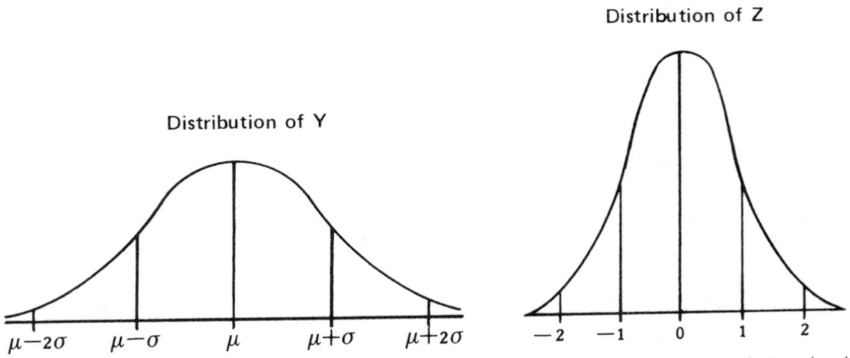

Figure 5–7. Normal density function of the random variable Y, with mean μ and standard deviation σ, and of the standardized random variable Z, with mean 0 and standard deviation 1.

We can also see in Table A2–1 that $F(-1.96) = 0.025$ and $F(1.96) = 0.975$, so that

$$F(1.96) - F(-1.96) = 0.975 - 0.025 = 0.95$$

Thus we see that 95% of a standard normal population lies between −1.96 and +1.96; however, throughout this book we shall often say, as an approximation, that 95% lies between −2 and +2.

We shall now see how Table A2–1 can be used to find the percentiles of *any* normal distribution. Consider, for example, the normal distribution illustrated in Figure 5–5(*b*), for which $\mu = 8$ and $\sigma = 2$. What proportion of the population lies below 12? To answer this question, we simply standardize the 12 and look the result up in Table A2–1. Thus, we set

$$z = \frac{12 - \mu}{\sigma} = \frac{12 - 8}{2} = 2$$

and Table A2–1 shows that $F(2) = 0.9772$, or approximately 97.5%. Following this approach, if we know the mean and standard deviation of a random variable and we know that its distribution is a member of the family of normal distributions, we can always standardize the random variable and use a standard table to make probability statements about it. We shall be using the cumulative standard normal distribution for this purpose, as well as other cumulative distributions, in later chapters.

SUMMARY

1. Traits are called variables for two main reasons: (a) Measurement error causes the values we observe to vary when measured repeatedly under the same conditions; and (b) all biological systems are dynamic and hence vary inherently. A random variable (response variable, variate) is just a variable observed as part of an experiment, or a well-defined sampling process, so that it can be associated with a probability distribution.

2. The distribution of a random variable is a description of the way the probability is distributed among the possible values the random variable can take on. There are two ways a distribution can be described: by a probability function (a probability density function in the case of a continuous random variable), or by a cumulative probability distri-

bution function. These each provide the same information about a random variable, but in a different way. A probability function is a rule that gives the probabilities (or densities) associated with different values of the random variable. A cumulative distribution function gives cumulative probabilities.

3. The binomial distribution is relevant for situations in which there are n observations of a dichotomous random variable, the n values arising independently of one another and each observation having the same probability of falling into a given category of the dichotomy; the total number of observations of one type is then binomially distributed.

4. The Poisson distribution can often be used as a model for random variables such as counts of rare events. It is a limiting case of the binomial distribution as the number of observations n becomes indefinitely large.

5. The uniform distribution (density) has a rectangular shape and is characterized by the fact that within well-defined limits, the probability of observing a value of the random variable in any interval is the same as that for any other interval of the same length.

6. The normal distribution (density) is bell-shaped and has a single peak. It is symmetric about its mean μ. The probability that a normally distributed random variable lies within one standard deviation of the mean is 68%; within two standard deviations, 95%.

7. A cumulative distribution gives the probability that a random variable is less than or equal to a particular value. The cumulative normal distribution has the shape of a sloping, elongated "S."

8. A standardized random variable has mean 0 and standard deviation 1. It is obtained by subtracting from a random variable its mean and dividing the result by its standard deviation. For the standard normal distribution, about 68% of the population lies between -1 and $+1$ and 95% between -2 and $+2$.

FURTHER READING

Tsokos, C.P. (1972) Probability Distributions: An Introduction to Probability Theory with Applications. Duxbury Press, Duxbury, Mass. (This book provides a mathematical description of many of the distributions likely to be encountered in a variety of applied problems. It is a good starting place for the reader who would like to know more about the mathematical aspects of these distributions.)

PROBLEMS

1. A variable (often denoted by a capital letter such as Y) that is observed as part of an experiment or a well-defined sampling process is called a
   A. binomial variable
   B. standardized variable
   C. random variable
   D. normalized variable
   E. uniform variable

2. A physician sent a sample of a patient's blood to a lipid laboratory for cholesterol and triglyceride determinations. The triglyceride determination was well within the normal range, but the cholesterol reading was 280 mg/dl. Being unconvinced that the patient had elevated cholesterol, the physician sent a second sample to the laboratory and the reading was 220. The disagreement between the two readings could be an example of
   A. the Poisson distribution and biological variability
   B. the binomial distribution and observer error
   C. the cumulative distribution and biological variability
   D. the random variable and the binomial distribution
   E. observer error and biological variability

3. If we conduct n trials, the outcome of each of which is either a success (with probability p) or failure (with probability 1 − p), the distribution of r, the total number of successes, is written

$$P(r) = \frac{n!}{(n - r)!r!} p^r(1 - p)^{n-r}$$

   This is known as
   A. the binomial distribution with parameters r and p
   B. the binomial distribution with parameters *n* and p
   C. the binomial distribution with parameters *n* and r
   D. the binomial distribution with probability p
   E. the binomial distribution with the mean *n*p

4. If there is a 2% chance that a child will be born with a congenital anomaly, what is the probability that no congenital anomaly will be found among four random births?

A. 0.92
B. 0.8
C. 0.2
D. 0.02
E. 0.08

5. A couple is at risk of having children with a recessive disease, there being a probability of 0.25 that each child is affected. What is the probability, if they have three children, that at least one will be affected?

A. 0.75
B. 0.02
C. 0.58
D. 0.42
E. 0.25

6. To determine whether mutations occur independently and with equal probability, a researcher sets up an experiment in which the number of mutant bacteria that appear in a certain volume of cell suspension can be counted as the number of colonies on an agar plate. One hundred cell suspensions are plated, and on each of the 100 plates the number of mutant colonies ranges from 0 to 9. In view of the purpose of the study, the distribution of these numbers should be compared to a

A. binomial distribution
B. Poisson distribution
C. uniform distribution
D. normal distribution
E. none of the above

7. A random variable, which we denote Y, is known to have a uniform distribution with probability density as follows:

$$f(y) = \frac{1}{10} \qquad 0 \leq y \leq 10$$

The probability an observed value of the random variable is between 4 and 6 is

A. $1/10$
B. $2/10$
C. $4/10$
D. $8/10$
E. 1

8. All the following are characteristics of the family of normal distributions except

A. positively skewed
B. mean equal to median
C. median equal to mode

D. mean equal to mode
E. symmetric

9. The normal distribution has two points of inflection. If the total area is 100%, what is the area under the normal curve between the two points of inflection?
   A. 99%
   B. 95%
   C. 90%
   D. 68%
   E. 50%

10. A normal distribution has mean 15 and standard deviation 3. What interval includes about 95% of the probability?
    A. 12–18
    B. 9–21
    C. 6–24
    D. 3–27
    E. none of the above

11. Weight is approximately normally distributed with mean 150 lb and standard deviation 10 lb. Which of the following intervals includes approximately two thirds of all weights?
    A. 145–155 lb
    B. 140–160 lb
    C. 130–170 lb
    D. 130–150 lb
    E. 150–170 lb

12. A random variable, which we denote Y, is known to be normally distributed with mean 100 and standard deviation 10. The probability that an observed value of this random variable is less than 90 or greater than 110 is approximately
    A. 0.15
    B. 0.32
    C. 0.68
    D. 0.84
    E. 0.95

13. A random variable, which we denote Y, is known to be normally distributed and to have mean 50 and standard deviation 5. What is the probability that the value of Y lies between 44 and 56? (Hint: Use Table A2–1.)
    A. 0.75
    B. 0.16
    C. 0.05

D. 0.87
E. 0.68

14. A random variable, which we denote Z, is known to have mean 0 and standard deviation 1. The random variable $Y = 10 + 2Z$ therefore has
  A. mean 0, standard deviation 2
  B. mean 10, standard deviation 0
  C. mean 0, standard deviation 4
  D. mean 10, standard deviation 2
  E. mean 2, standard deviation 10

15. A cumulative distribution
  A. expresses the same information as a probability or density function, but in a different way
  B. states the probability that a random variable is less than or equal to a particular value
  C. always takes on a value between zero and one
  D. all of the above
  E. none of the above

CHAPTER SIX

Key Concepts

estimate, estimator
standard error (of the mean)
unbiased, biased
minimum variance unbiased estimator, efficient estimator, robust estimator
maximum likelihood estimate
normal range, confidence limits or confidence interval
(Student's) t-distribution, degrees of freedom
pooled variance

Symbols and Abbreviations

d.f.	degrees of freedom
s	sample standard deviation (estimate)
S	sample standard deviation (estimator)
s^2	sample variance (estimate)
S^2	sample variance (estimator)
s.e.m.	standard error of the mean
t	percentile of Student's t-distribution or corresponding test statistic
x, y	particular values of (random) variables
X, Y	random variables
$\bar{y}$	sample mean of y (estimate)
$\bar{Y}$	sample mean of Y (estimator)
z	particular value of a standardized normal random variable
Z	standardized normal random variable
π	population proportion
μ	population mean
σ	population standard deviation
σ^2	population variance

Estimates and Confidence Limits

ESTIMATES AND ESTIMATORS

It will be helpful at this point to distinguish between an estimate and an estimator. An **estimator** is a rule that tells us how to determine from any sample a numerical value to estimate a certain population parameter, whereas an **estimate** is the actual numerical value obtained from a particular sample. Suppose we select a random sample of 10 students from a class of 200, measure their heights, and find the sample mean. If we consider the "sample mean" to be the rule that tells us to add up the 10 heights and divide by 10, it is an estimator, and we shall denote it $\overline{Y}$. If, on the other hand, we consider the "sample mean" to be the result we obtain from a particular sample—say 70 inches—it is an estimate and we shall denote it $\overline{y}$. The estimator $\overline{Y}$ is a random variable that takes on different values from sample to sample. The estimate $\overline{y}$, on the other hand, is the numerical value obtained from one particular sample. We say the sample mean $\overline{Y}$ is an estimator, and $\overline{y}$ an estimate, of the population mean μ. Thus, particular values of an estimator are estimates. Different samples might yield $\overline{y} = 69$ inches or $\overline{y} = 72$ inches as estimates of the mean of the entire class of 200 students, but the estimator $\overline{Y}$ is always the same: add up the 10 heights and divide by 10.

Note we have retained our convention of using capital letters to denote random variables (estimators) and lowercase letters to denote specific values of random variables (estimates). The word "estimate" is often used in both senses, and there is no harm in this provided you understand the difference. To help you appreciate the difference, however, we shall be careful to use the two different words and symbols, as appropriate.

Since an estimator is a random variable, and hence has a distribution, it is of interest to determine its variance and standard deviation. By selecting many different samples and obtaining the corresponding estimate for each sample, we obtain a set of data (estimates) from which we can compute a variance and standard deviation in the usual way. The problem with this approach is that it requires that many samples be studied and is therefore not practical. Fortunately, however, there is a way of estimating the variance and standard deviation of an estimator from the information available in a single sample. If the population variance of a random variable is σ^2, it can be proved mathematically that the sample mean $\bar{y}$ of n independent observations has variance equal to σ^2/n and standard deviation equal to $\sigma/\sqrt{n}$. We can estimate these quantities by substituting s for σ, where s is computed from a single sample in the usual way. We see immediately from the formula σ^2/n that the larger the sample size (n), the smaller is the variance and standard deviation of the sample mean. In other words, the sample mean calculated from large samples varies less, from sample to sample, than the sample mean calculated from small samples.

It is usual practice to call the standard deviation of an estimator the standard error (often abbreviated s.e.) of that estimator. The standard deviation of the sample mean $(\sigma \sqrt{n})$ is, therefore, often called the **standard error of the mean.** When applied to the sample mean, there is no real difference between the terms "standard deviation" and "standard error," but the latter is often used when we are referring to variability due to error, as opposed to natural variability. The fact that men have variable heights is in no way due to "error," but the fact that our estimate of mean height differs from the true mean can be considered an error. Furthermore, from now on in this book, as in so much of the medical literature, we shall use the term "standard error" to indicate the *estimated* standard deviation of the estimator. Thus, the standard error of the mean (often abbreviated s.e.m.) is $s/\sqrt{n}$.

NOTATION FOR POPULATION PARAMETERS, SAMPLE ESTIMATES, AND SAMPLE ESTIMATORS

We have already indicated that we denote population parameters by Greek letters and sample statistics by Roman letters. The former are always fixed constants, whereas the latter, in the context of repeated sampling, can be considered random variables. The particular estimates calculated from one sample (such as 69 inches or 72 inches in the above example) are fixed constant quantities, but they can be viewed as particular outcomes of random variables in the context of examining many, many samples. Depending on how they are viewed, we use lowercase or uppercase Roman letters. Thus, assuming a sample of size n, our notation is as follows:

Name	Parameter	Estimate	Estimator
Mean	μ	$\bar{y}$	$\bar{Y}$
Variance	σ^2, σ_Y^2	s^2, s_Y^2	S^2, S_Y^2
Standard deviation	σ, σ_Y	s, s_Y	S, S_Y
Standard deviation of the mean	$\sigma_{\bar{Y}} = \sigma_Y/\sqrt{n}$	$s_{\bar{Y}} = s_Y/\sqrt{n}$	$S_{\bar{Y}} = S_Y/\sqrt{n}$

Thus μ is the population mean (parameter), $\bar{y}$ the sample mean (estimate). Similarly, σ is the population standard deviation, s the sample standard deviation; in other words, σ is the standard deviation of a random variable Y, and s is the estimate of σ we calculate from a sample. When we want to stress the fact that the relevant random variable is Y, we write σ_Y or s_Y—the standard deviation of Y. Analogously, $\sigma_{\bar{Y}}$ is the standard deviation of $\bar{Y}$, the standard deviation of the sample mean of n observations, while $s_{\bar{y}}$ is the sample estimate of this quantity, which, as noted earlier, is often called the **standard error of the mean.** Thus $\sigma_{\bar{Y}} = \sigma_Y/\sqrt{n}$ and $s_{\bar{Y}} = s_Y/\sqrt{n}$.

PROPERTIES OF ESTIMATORS

We usually **estimate** the population mean by the sample mean; and we often **estimate** the population variance by the sample variance. Why do we do this? Why not, for example, estimate the population mean by the average of the smallest and largest values in our sample? The answer is that we choose estimators, and hence estimates, that have certain "good" properties. An estimator is a random variable and has a distribution. The characteristics of this distribution determine the goodness of the estimator. In the previous example, suppose we take all possible samples of 10 students from the class of 200 and compute the mean height from each sample. These sample means vary in the sense that they have different values from sample to sample, and if we average them, we obtain a "mean of means," which is equal to the population mean—the mean height from all 200 students. In a case such as this, in which the average of the sample estimates for all possible samples equals the value of the parameter being estimated, we say the estimator is **unbiased:** the mean of the estimator's distribution is equal to the parameter being estimated. When this does not occur, the estimator is **biased.** In general, when we have a random sample, the sample mean is an unbiased estimator of the population mean and the sample variance is an unbiased estimator of the population variance; but the sample standard deviation is not an unbiased estimator of the population standard deviation.

Recall that we use $n - 1$ rather than n as a divisor when we average the sum of squared deviations from the sample mean in computing the variance. If we used n as a divisor, we would find that the average of all possible estimates is equal to $(n - 1)/n$ times the population variance and is thus biased; such an estimator leads to an estimate that is, on an average, smaller than the population value. With $n - 1$ as a divisor, we have an unbiased estimator of the variance.

We should also like our estimate to be close to the parameter being estimated as often as possible. It is not very helpful for an estimate to be correct "on an average" if it fluctuates widely from sample to sample. Thus, for an estimator to be good, its distribution should be concentrated fairly closely about the true value of the parameter of interest; in other words, the variance of the estimator's distribution should be small. If we have a choice among several estimators that are competing, so to speak, for the job of estimating a parameter, we might proceed by eliminating any that are biased and then, from among those that are unbiased, choose the one with the smallest variance. Such an estimator is called a **minimum variance unbiased estimator,** or an **efficient estimator.** It can be shown mathematically that if the underlying population is normally distributed, the sample mean and the sample variance are minimum variance unbiased estimators of the population mean and the population variance, respectively. In other situations, however, the sample descriptive statistic may not have this property. The sample mean is a minimum variance unbiased estimator of the population mean when the underlying population is binomially or Poisson-distributed, but in these two cases an analogous statement cannot be made about the sample variance. In fact, the efficient estimator of the variance of a Poisson distribution is provided by the sample mean!

We have discussed only two important properties of good estimators. Other properties are also of interest, but the important thing is to realize that criteria exist for evaluating estimators. One estimator may be preferred in one situation and a second, competing estimator may be preferred in another situation. There is a tendency to use estimators whose good properties depend on the data following a normal distribution, even though the data are not normally distributed. This is unfortunate, since more appropriate methods of analysis are available. You cannot hope to learn all of the considerations that must be made in choosing estimators that are appropriate for each situation that might arise, but you should be aware of the necessity of examining such issues.

An estimator that usually has good properties, whatever the situation, is called **robust.** It is to our advantage if we can find estimators that are robust. The sample mean is a robust estimator of the population mean.

MAXIMUM LIKELIHOOD

Having accepted the fact that estimators should have certain desirable properties, how do we find an estimator with such properties in the first place? Again there are many ways of doing this, and no single approach is preferable in every situation. A full discussion of these methods requires mathematical details that are beyond the scope of this book, but we must describe two very important approaches to deriving estimators, since they are frequently mentioned in the literature. One is based on a method known as maximum likelihood estimation, and the other on a method called least squares estimation. The method of least squares will be discussed in Chapter 9. **Maximum likelihood estimates** of parameters such as the population mean and variance are those values of the parameters that make the probability, or likelihood, of our sample as large as possible—a maximum. Intuitively, they are those parameter values that make the data we observe "most likely" to occur in the sample.

A simple example will illustrate the principle. Suppose we wish to estimate the proportion of a population that is male. We take a random sample of n persons from the population and observe in the sample y males and $n - \mathrm{y}$ females. Suppose now the true population proportion is π, so that each person in the sample has a probability π of being male. What then is the probability, or likelihood, of our sample, which contains y males and $n - \mathrm{y}$ females? We learned in Chapter 5 that the probability of this happening is given by the binomial distribution, with parameters n and π; i.e.,

$$\mathrm{P(y\ males)} = \frac{n!}{\mathrm{y}!(n - \mathrm{y})!}\,\pi^{\mathrm{y}}(1 - \pi)^{n-\mathrm{y}}$$

Fixing y, which is known, and considering π to be a variable, we now ask: What value of π makes this probability the largest—a maximum? We shall call that value the maximum likelihood estimate of π. We shall not prove it here, but it can be shown that in this case the likelihood is largest when we set π equal to y/n, the sample proportion of male persons. [It is instructive to take the time to verify this numerically. Suppose, for example, $n = 5$ and $y = 2$. Calculate P(y male) for various values of π (e.g., $\pi = 0.2, 0.3, 0.4, 0.5$). You will find that it is largest when $\pi = 0.4$.] Thus, the maximum likelihood estimate of π in this situation is y/n. Analogously, the maximum likelihood estimator of π is Y/n, where Y is the random variable denoting the number of male persons we find in samples of size n from the population.

Except as noted in Appendix 1, maximum likelihood estimators have the following important properties for samples comprising a very large number of study units; i.e., as the number of study units tends to infinity, maximum likelihood estimators have the following (so-called asymptotic) properties:

1. They are unbiased: the mean of the estimator (i.e., the mean of many, many estimates) will equal the true value of the parameter being estimated.
2. They are efficient: the variance of the estimator (i.e., the variance of many, many estimates) is the smallest possible for any asymptotically unbiased estimator.
3. The estimators are normally distributed. The utility of this last property will become apparent later in this chapter.

In certain instances, maximum likelihood estimators have some of these properties for all sample sizes n; in general, however, they have these properties only for very large sample sizes. How large n must be for these properties to be enjoyed, at least approximately, depends on the particular situation, and is often unknown. Other problems can occur with maximum likelihood estimators: we must know the mathematical formula for the distribution, they may be difficult to compute, they may not exist, and if they do exist they may not be unique. Nevertheless, the principle of maximum likelihood estimation has a great deal of intuitive appeal and it is widely used.

ESTIMATING INTERVALS

In the literature, estimates are often given in the form of a number plus or minus another number. For example, the mean serum cholesterol level for a group of 100 medical students might be reported as 186 $\pm$ 32 mg/dl. Unfortunately, these numbers are often reported without any explanation of what the investigator is attempting to estimate. It is a standard convention for the first number (186) to be the parameter estimate, but there is no standard convention regarding what the number after the $\pm$ sign represents. In some instances, the second number is a simple multiple (usually 1 or 2) of the estimated standard deviation of the random variable of interest. In other instances, it may be some multiple of the standard error of the estimator. In our example, 186 $\pm$ 32 mg/dl represents the mean serum cholesterol level plus or minus one standard deviation, estimated from a group of 100 medical students. If we divide 32 mg/dl by the square root of the sample size (i.e., by 10), we obtain the standard error of the mean, equal to 3.2 mg/dl. Thus the interval 186 $\pm$ 3.2 mg/dl represents the mean serum cholesterol level plus or minus one standard error of the mean. Clearly it is important to know which is being quoted, and to know whether the number after the $\pm$ sign is once, twice or some other multiple of the estimated standard deviation or standard error of the mean. We now turn to a discussion of how the different intervals should be interpreted.

We shall see later in this chapter that the standard error of the mean is used to define an interval that we believe contains the true value of the mean. For now, suppose we wish to estimate an interval that includes most of the population values. Suppose that our population of cholesterol values is normally distributed, with mean μ and standard deviation σ. Then we know that approximately two thirds of the population lies between $\mu - \sigma$ and $\mu + \sigma$, i.e., in the interval $\mu \pm \sigma$. We can estimate this interval by $\bar{y} \pm s$ (i.e., 186 $\pm$ 32 mg/dl). Thus, provided cholesterol levels are normally distributed and our sample size is large enough for the estimates to be close to the true parameters, we can expect about two thirds of the population to have cholesterol levels between 154 and 218 mg/dl. Similarly, we know that about 95% of the population lies in the interval $\mu \pm 2\sigma$, which for our example is estimated by 186 $\pm$ 64. Thus we might estimate that 95% of the population cholesterol levels are between 122 and 250 mg/dl.

Such intervals, or limits, are often quoted in the literature in an attempt to define a **normal range** for some quantity. They must, however, be very cautiously interpreted on two counts. First, the assumption of normality is not a trivial one. Suppose it was quoted, on the basis of the data in Table 3–1, that $\bar{y} \pm 2s$ for triglyceride levels in male medical students is 111 ± 118 mg/dl ($\bar{y} = 111$, $s = 59$). Should we expect 95% of the population to be in this interval? Note that this interval is from −7 to 229 mg/dl. Triglyceride levels are not at all normally distributed, but follow a very positively skewed distribution in the population (see Figs. 3–2 and 3–4); this explains why $\bar{y} - 2s$ is negative and hence an impossible quantity. We must not assume that $\mu \pm 2\sigma$ includes about 95% of the population for a nonnormally distributed random variable. Second, there is an implicit assumption (in addition to normality) that the sample estimates are close to the true parameter values. It is rarely realized that it takes very large sample sizes for this assumption to be adequately met. If the sample size is less than 50, then $\bar{y} \pm 2s$ includes, on an average, less than 95% of a normally distributed population. When the sample size is 10, it includes on an average only 92%, and when the sample size is down to 8, it includes on an average less than 90%. Although these consequences might be considered small enough to neglect, it must be realized that these percentages are only averages. The results from any one sample could well include less than 75% of the population, as can be seen from Table 6–1. We see from this table, for example, that when the sample size is 10, there is a 0.12 probability that $\bar{y} \pm 2s$ includes less than 75% of the population. Even when the sample size is 100, there is still a 0.06 probability that $\bar{y} \pm 2s$ includes less than 75% of the population. Thus, if cholesterol levels are

TABLE 6–1. **Approximate Proportion of Samples from a Normal Distribution in Which the Estimated Mean ± 2 s.d. Will Fail to Include the Indicated Percentage of the Population**

	Percentage of the Population		
Sample Size	**75**	**90**	**95**
10	0.12	0.18	0.24
20	0.09	0.12	0.15
30	0.08	0.10	0.12
40	0.07	0.09	0.11
50	0.07	0.09	0.11
100	0.06	0.07	0.08

normally distributed, the interval 122 to 250 mg/dl has a small but non-negligible probability (0.06) of including less than 75% of the population.

DISTRIBUTION OF THE SAMPLE MEAN

We have seen that if Y is normally distributed, we can write, approximately,

$$P(\mu - 2\sigma \leq Y \leq \mu + 2\sigma) = 0.95$$

and

$$P\left(-2 \leq \frac{Y - \mu}{\sigma} \leq 2\right) = 0.95$$

The 97.5th percentile of the standard normal distribution is about 2, and the 2.5th percentile is about -2. Now consider $\overline{Y}$, the mean of n such normally distributed random variables. We have already learned that $\overline{Y}$ has mean μ and standard deviation $\sigma/\sqrt{n}$. Furthermore, it is normally distributed. As an example, suppose we have a population in which the mean height is 67 inches and the standard deviation is 3 inches. We take samples of four persons each and average their heights. Then the distribution of these averages is normal with mean 67 inches and standard deviation $3/\sqrt{4}$ inches $=$ 1.5 inches. This is pictured in Figure 6–1. Notice that whereas about 95% of the population lies in the interval 61 to 73 inches, about 95% of the means lie in the interval 64 to 70 inches.

Now let us subtract the mean from $\overline{Y}$ and divide the difference by *its* standard deviation, i.e.,

$$Z = \frac{\overline{Y} - \mu}{\sigma_{\overline{Y}}}$$

We denote the result Z, because that is the letter usually used for a standardized normal random variable; with about 95% probability it lies between -2 and $+2$. Thus we can write

$$P(-2 \leq Z \leq 2) \doteq 0.95$$

or

$$P\left(-2 \leq \frac{\overline{Y} - \mu}{\sigma_{\overline{Y}}} \leq 2\right) \doteq 0.95$$

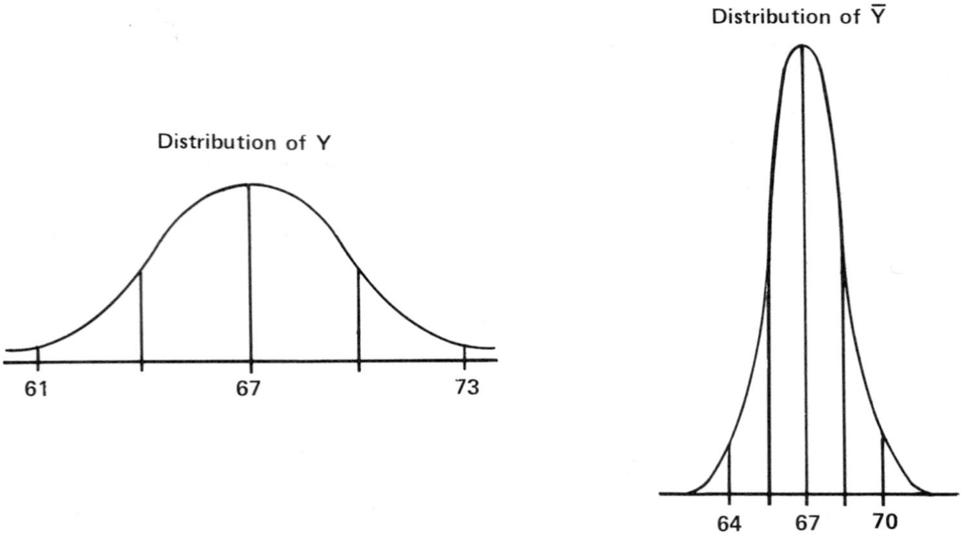

Figure 6–1. Normal density function of a person's height, Y, with mean 67 inches and standard deviation 3 inches, and of the average of a sample of four persons' heights, $\overline{Y}$, with the same mean 67 inches but standard deviation 1.5 inches.

which is equivalent to

$$P(\overline{Y} - 2\sigma_{\overline{Y}} \leq \mu \leq \overline{Y} + 2\sigma_{\overline{Y}}) \doteq 0.95$$

(The derivation of this equivalence is given in Appendix 1.) Remember, $\sigma_{\overline{Y}} = \sigma_Y/\sqrt{n}$ for a sample of size n.

Let us now summarize the various facts we have learned about the sample mean $\overline{Y}$:

1. The mean of $\overline{Y}$ is μ.

2. The variance of $\overline{Y}$ is σ^2/n (i.e., $\sigma^2_{\overline{Y}} = \sigma^2_Y/n$).

3. The standard deviation of $\overline{Y}$ is $\sigma/\sqrt{n}$ (i.e., $\sigma_{\overline{Y}} = \sigma_Y/\sqrt{n}$).

4. $\overline{Y}$ is normally distributed. This is strictly true only if Y is normally distributed. It is a remarkable fact, however, that it also tends to be true, for moderate to very large sample sizes, almost irrespective of how Y is distributed (the word "almost" is added to allow for some special situations that, although mathematically possible, do not usually occur in practice). Usually, a mean of five or more observations is for all intents and purposes normally distributed.

5. It therefore follows, provided $\overline{Y}$ is based on five or more observations, that

$$P\left(\overline{Y} - \frac{2\sigma_Y}{\sqrt{n}} \leq \mu \leq \overline{Y} + \frac{2\sigma_Y}{\sqrt{n}}\right) \doteq 0.95$$

CONFIDENCE LIMITS

Consider once more the triglyceride levels in Table 3–1, for which $\overline{y} = 111$ mg/dl, $s = 59$ mg/dl, and $n = 30$. The standard error of the mean is $59/\sqrt{30} = 11$ mg/dl. Let us assume for the moment that the true population values are $\sigma_Y = 59$ and $\sigma_{\overline{Y}} = 11$ mg/dl. (We shall see how to avoid this assumption later.) Thus the interval $\overline{y} \pm 2\sigma_{\overline{Y}}$ is 111 ± 22, or 89 to 133 mg/dl. Can we therefore say that there is about a 95% probability that μ lies in this interval? In other words, can we write

$$P(89 \leq \mu \leq 133) \doteq 0.95?$$

Regardless of how good the approximation might be, such an expression is impossible. The true mean μ is a fixed quantity, not a random variable, and the fact that we do not know it does not alter this. Similarly, 89 and 133 are fixed quantities. Either μ lies between 89 and 133 mg/dl, or it does not. We cannot talk about the "probability" of this being the case, since probability is a property associated with random variables. Thus it is meaningful to write

$$P(\overline{Y} - 2\sigma_{\overline{Y}} \leq \mu \leq \overline{Y} + 2\sigma_{\overline{Y}}) \doteq 0.95$$

but not to write

$$P(\overline{y} - 2\sigma_{\overline{Y}} \leq \mu \leq \overline{y} + 2\sigma_{\overline{Y}}) \doteq 0.95$$

Now you can see why we have taken pains to distinguish between $\overline{Y}$ and $\overline{y}$. The former is a random variable, about which we can make a probability statement; the latter is a fixed quantity, about which we cannot make a probability statement.

Despite the fact that we cannot talk about the probability of μ lying between 89 and 133 mg/dl, this interval is clearly somehow related to the unknown mean μ, and we should be reasonably certain (perhaps "95% certain") that μ is in fact between 89 and 133 mg/dl. We call these numbers the 95% **confidence limits, or confidence interval,** for μ, and we say that μ lies between 89 and 133 mg/dl with 95% confidence.

Thus $\bar{y} \pm 2\sigma_{\bar{Y}}$ (i.e., $\bar{y} \pm 2\sigma_Y/\sqrt{n}$) gives an approximate 95% confidence interval for the mean. This is strictly true only if Y is normally distributed, but tends to be true, for moderately large or large samples, whatever the distribution of Y. A confidence interval is to be interpreted as follows: if we were to find many such intervals, each from a different sample but in exactly the same fashion, then in the long run about 95% of our intervals would include the true mean and 5% would not. We cannot say that there is a 95% probability that the true mean lies between the two values we obtain from a particular sample, but we can say that we have 95% confidence that it does so. We know that if we estimate μ by a single number $\bar{y}$, we cannot expect to be so lucky as to have $\bar{y} = \mu$. With a single number as our estimate, we have no feel for how far off we might be with our estimate. By using a confidence interval to estimate μ, we have a range of values that we think, with some degree of confidence, contains the true value μ. We shall now briefly indicate how confidence limits are calculated in several specific situations.

CONFIDENCE LIMITS FOR A PROPORTION

Suppose we wish to estimate the probability a newborn is male. We take a random sample of 1,000 births and find 526 are male. We therefore estimate the proportion of births that are male as being 0.526. What would the 95% confidence limits be? Since 0.526 is a maximum likelihood estimate based on a large sample, it can be considered as the outcome of a normally distributed random variable. Now if Y follows a binomial distribution with parameters n and π, then it can be shown that the variance of Y/n (y/n is 526/1,000 in our particular sample) is $\pi(1 - \pi)/n$. Thus Y/n is about normally distributed with mean π and standard deviation $\sqrt{\pi(1 - \pi)/n}$, and so an approximate 95% confidence interval for π is

$$\frac{y}{n} \pm 2\sqrt{\frac{\pi(1 - \pi)}{n}}$$

which we estimate by

$$0.526 \pm 2\sqrt{\frac{0.526(1 - 0.526)}{1000}} = 0.526 \pm 0.032$$

Thus we have about 95% confidence that the true proportion lies between 0.494 and 0.558.

Notice that to calculate this interval we substituted our estimate, 0.526, for π. For a large sample this is adequate. Had the sample been small, we should not assume Y/n is normally distributed, and it would have been necessary to use special tables that have been calculated from the binomial distribution. As a rule of thumb, the approximation is adequate provided both $n\pi$ and $n(1 - \pi)$ are greater than 5. Note also that because the estimator is about normally distributed, in large samples a 95% confidence interval can always be obtained from a maximum likelihood estimate by adding and subtracting twice the standard error of the estimator.

CONFIDENCE LIMITS FOR A MEAN

Consider our example in which we determined 95% confidence limits for mean triglyceride level as being 89 to 133 mg/dl. Recall that we assumed we knew $\sigma_Y = 59$ and $\sigma_{\bar{Y}} = 11$ mg/dl, whereas in fact these were really the sample estimates s_Y and $s_{\bar{Y}}$, respectively. Usually we do not know the true standard deviation, σ_Y, so to be of any practical use we must be able to calculate confidence limits without it. Does it make any difference if we simply substitute $s_{\bar{Y}}$ for $\sigma_{\bar{Y}}$? The answer is that it does not, for all practical purposes, if the sample size is more than 30. When the sample size is smaller than this, we must allow for the fact that we do not know $\sigma_{\bar{Y}}$, as follows.

We have seen that if Y is normally distributed, then $(\bar{Y} - \mu)/\sigma_{\bar{Y}}$ follows a standard normal distribution for which the approximate 2.5th and 97.5th percentiles are, respectively, -2 and $+2$. Analogously, substituting $S_{\bar{Y}}$ for $\sigma_{\bar{Y}}$, $(\bar{Y} - \mu)/S_{\bar{Y}}$ follows a distribution called **Student's t-distribution** with $n - 1$ degrees of freedom. Like the standard normal distribution, this is symmetric about zero, so that the 2.5th percentile is simply the negative of the 97.5th percentile. Denote the 97.5th percentile $t_{97.5}$. Then the 95% confidence limits are $\bar{y} \pm t_{97.5}s_{\bar{Y}}$.

The distribution of $(\bar{Y} - \mu)/S_{\bar{Y}}$ was derived by a mathematician who worked for the Guinness Brewing Company in Ireland and who published his statistical papers under the pseudonym "Student." This quantity is denoted t, and hence the name "t-distribution." Just as the normal distribution is a family of distributions, a particular one being determined by the parameters μ and σ, so is the t-distribution a family of distributions—but in this case the particular distribution is determined by a parameter known as the "number of degrees of freedom," a concept we shall explain shortly. Each t-distribution is similar to the standard normal distribution but has thicker tails, as illustrated in Figure 6–2. The fewer the degrees of freedom, the thicker the tails are. As the number of degrees of freedom becomes indefinitely large (in which case $\sigma = s$), the t-distribution becomes the same as the standard normal distribution. This can be seen in the following 97.5th percentiles, abstracted from Table 2 in Appendix 2:

1 degree of freedom	12.706
10 : : :	2.228
30 : : :	2.042
(Standard normal) ∞ : : :	1.960

Now you can see the basis for saying that for all practical purposes it makes no difference if we substitute $s_{\bar{Y}}$ for $\sigma_{\bar{Y}}$ when the sample size is more than 30; 2 is almost as close an approximation to 2.042 ($t_{97.5}$ when there are 30 degrees of freedom, corresponding to $n = 31$) as it is to 1.960 (the 97.5th percentile of the standard normal distribution).

Now let us go back once more to our example of triglyceride levels (from the data in Table 3–1), for which $\bar{y} = 111$ mg/dl, $s_{\bar{Y}} = 59$ mg/dl, and $n = 30$. Calculating with more accuracy than previously, we find $s_{\bar{Y}} = 59/\sqrt{30} = 10.77$ mg/dl, and from Table 2 in Appendix 2 we find that for $n - 1 = 29$ degrees of freedom, $t_{97.5} = 2.045$. From this we find that $\bar{y} \pm t_{97.5}s_{\bar{Y}}$ is $111 \pm (2.045)(10.77) = 111 + 22.0$ mg/dl. Thus in this case our earlier approximation was adequate. Had the sample size been much smaller (e.g., 10 or less), however, the approximation $\bar{y} \pm 2s_{\bar{Y}}$ would have led to an interval that is much too short.

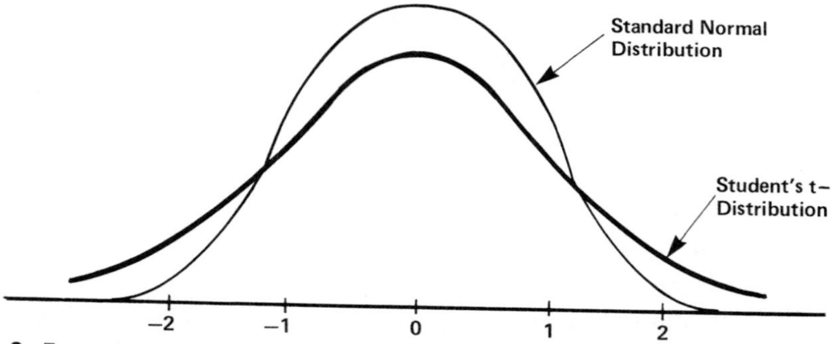

Figure 6–2. Example of the density function of Student's t-distribution compared to that of the standard normal distribution.

THE CONCEPT OF DEGREES OF FREEDOM

The term **degrees of freedom,** abbreviated d.f., will occur again and again, not only in connection with the t-distribution, but also in many other problems. Basically, the degrees of freedom refer to the number of "independent" observations in a quantity. You know that the sum of the angles in a triangle is equal to 180°. Suppose you were asked to choose the three angles of a triangle. You have two degrees of freedom in the sense that you may choose two of the angles, but then the other is automatically determined because of the restriction that the sum of the three is 180°. Suppose you are asked to choose three numbers with no restrictions on them. You have complete freedom of choice in specifying all three numbers and hence, in that case, you have three degrees of freedom.

Now suppose you are asked to choose six numbers (which we shall call y_1, y_2, y_3, y_4, y_5, and y_6) such that the sum of the first two is 16 and also such that the sum of all of them is 40. There are six numbers to be specified, but you do not have freedom of choice for all six. You have to take into account the restrictions

$$y_1 + y_2 = 16$$

and

$$y_1 + y_2 + y_3 + y_4 + y_5 + y_6 = 40$$

As soon as you select y_1, then $y_2 = 16 - y_1$, and so y_2 is completely determined. Of the remaining numbers, $y_3 + y_4 + y_5 + y_6 = 40 - 16 = 24$. Thus only three of the numbers y_3, y_4, y_5, and y_6 can be freely chosen. If we choose y_3, y_4, and y_5, for example, y_6 is predetermined as follows:

$$y_6 = 24 - (y_3 + y_4 + y_5)$$

Hence, the total number of degrees of freedom in this example is $1 + 3 = 4$.

In computing a variance, we use as the divisor a number that makes the variance an unbiased estimator of the population variance. This divisor is the number of degrees of freedom associated with the estimator once an estimate of the mean has been made. Recall that our divisor for the variance is the size of the sample minus one (i.e., $n - 1$). Once the mean is fixed at its sample value, there are only $n - 1$ degrees of freedom associated with permissible values of the numbers used to compute the variance. This same number is also the number of degrees of freedom associated with the estimated standard deviation, and with the t-distribution used to obtain confidence limits for the mean from the formula $\bar{y} \pm ts_{\bar{Y}}$.

Other types of problems with a variety of restrictions and degrees of freedom are considered in subsequent chapters of this book. In every instance the number of degrees of freedom is associated with a particular statistic (such as $s_{\bar{Y}}$). It is also the appropriate value to use as the parameter of a distribution (such as the t-distribution) when using that statistic for a particular purpose (such as calculating confidence limits).

CONFIDENCE LIMITS FOR THE DIFFERENCE BETWEEN TWO MEANS

Suppose we compare two drugs, A and B, each aimed at lowering serum cholesterol levels. Drug A is administered to one group of patients (sample 1) and drug B to a second group (sample 2), with the patients randomly assigned to the two groups so that the samples are independent. If we use the estimators $\bar{Y}_1$ and $\bar{Y}_2$ to find estimates $\bar{y}_1$ and $\bar{y}_2$ of the post-treatment serum cholesterol means for drug A and drug B, respectively, we might want to construct a confidence interval for the true difference $\mu_1 - \mu_2$. In this situation a 95% confidence interval would be given by

$$\bar{y}_1 - \bar{y}_2 \pm t_{97.5}s_{\bar{Y}_1-\bar{Y}_2}$$

where $t_{97.5}$ is the 97.5th percentile of the t distribution, with degrees of freedom equal to that associated with $s_{\bar{Y}_1-\bar{Y}_2}$, the standard error of $\bar{Y}_1 - \bar{Y}_2$. If the two samples have the same true variance σ^2, then the respective sample variances s_1^2 and s_2^2 are both estimates of the same quantity σ^2. In such instances we can average or "pool" the sample variances to obtain a **pooled estimate** s_p^2 of σ^2:

$$s_p^2 = \frac{(n_1 - 1)s_1^2 + (n_2 - 1)s_2^2}{n_1 + n_2 - 2}$$

Note that when we pool s_1^2 and s_2^2, we weight each by the number of degrees of freedom associated with it. Note also that we obtain a (weighted) average of sample *variances,* not of standard deviations. We take the square root of this pooled variance to obtain the (sample) pooled standard deviation s_p. Next, you need to know that when we have two independent random variables, the variance of their difference is equal to the sum of their variances. Thus the variance of $\bar{Y}_1 - \bar{Y}_2$ is the variance of $\bar{Y}_1$

plus the variance of $\overline{Y}_2$, i.e.,

$$\sigma^2_{\overline{Y}_1-\overline{Y}_2} = \frac{\sigma^2}{n_1} + \frac{\sigma^2}{n_2} = \sigma^2\left(\frac{1}{n_1} + \frac{1}{n_2}\right)$$

and it follows that the standard deviation of $\overline{Y}_1 - \overline{Y}_2$ is the square root of this, or $\sigma\sqrt{1/n_1 + 1/n_2}$. The standard error of $\overline{Y}_1 - \overline{Y}_2$ is obtained by substituting the estimate s_p for σ in this expression, so that the confidence interval is

$$\overline{y}_1 - \overline{y}_2 \pm t_{97.5}s_{\overline{Y}_1-\overline{Y}_2} = \overline{y}_1 - \overline{y}_2 \pm t_{97.5}s_p\sqrt{\frac{1}{n_1} + \frac{1}{n_2}}$$

in which $t_{97.5}$ is the 97.5th percentile of the t distribution with $n_1 + n_2 - 2$ degrees of freedom. The number of degrees of freedom associated with s_p is the sum of the number of degrees of freedom associated with s_1 and the number associated with s_2. Once $\overline{y}_1$ and $\overline{y}_2$ are known, there are $n_1 + n_2 - 2$ independent observations used in computing s_p.

If the two samples do not have the same true variance, other methods, which we shall not detail here, must be used. The same method, however, leads to a good approximation of a 95% confidence interval even if the true variances are unequal, provided $n_1 = n_2$ (i.e., provided the two samples have the same size). When the two samples have different sizes, we must first determine whether it is reasonable to suppose that the two true variances are equal. This and other similar topics are the subjects of the next chapter.

SUMMARY

1. An estimator is a rule for calculating an estimate from a set of sample values. It is a random variable that takes on different values—estimates—from sample to sample. The mean of n independent random variables, each with variance σ^2, has variance σ^2/n and standard deviation $\sigma/\sqrt{n}$. The latter, or its estimate $s/\sqrt{n}$, is called the standard error of the mean.

2. An unbiased estimator is one whose mean is equal to the parameter being estimated. The mean and variance of a random sample are unbiased estimators of the population mean and variance. An efficient estimator is an unbiased estimator that has minimum variance. In the case of normally, binomially, or Poisson-distributed random variables, the sample mean is a minimum variance unbiased estimator. The sample mean is a robust estimator (i.e., a good estimator of the population mean in a wide variety of situations).

3. Maximum likelihood estimates are parameter values that make the likelihood (probability) of the sample a maximum. In large samples, maximum likelihood estimators are usually unbiased, efficient, and normally distributed. Sometimes they also have these properties in small samples.

4. The estimated mean ± 2 standard deviations is often calculated as a "normal range" that contains about 95% of the population values. There is no guarantee that this approximation is good unless the population is normally distributed and the interval is calculated from a large sample. Sample means, on the other hand, tend to be normally distributed regardless of the form of the distribution being sampled.

5. A parameter can be said to lie within a specified interval with a certain degree of confidence, not with any degree of probability. The estimator, but not the estimate, of a 95% confidence interval can be said to have 95% probability of including the parameter. A particular 95% confidence interval for the mean should be interpreted as follows: if many such intervals were to be calculated in the same fashion, each from a different sample, then in the long run 95% of such intervals would include the true mean.

6. For large samples, the maximum likelihood estimate of a parameter ± twice the standard error is an approximate 95% confidence interval for that parameter. For sample sizes larger than 30, the sample mean ± two standard errors of the mean provide approximate 95% confidence limits for the mean. For smaller samples and a normally distributed random variable, 95% confidence limits are given by $\bar{y} \pm t_{97.5}\, s/\sqrt{n}$, where $t_{97.5}$ is the 97.5th percentile of Student's t-distribution with $n - 1$ degrees of freedom.

7. If two independent samples come from populations with the same common variance, a pooled estimate, s_p^2, of the variance can be obtained by taking a weighted average of the two sample variances, s_1^2 with $n_1 - 1$ degrees of freedom and s_2^2 with $n_2 - 1$ degrees of free-

dom, weighting by the number of degrees of freedom. The pooled estimate then has $n_1 + n_2 - 2$ degrees of freedom and can be used to determine a 95% confidence interval for the difference between the two means: $\bar{y}_1 - \bar{y}_2 \pm t_{97.5} s_p \sqrt{1/n_1 + 1/n_2}$. This same interval is about correct even if the two variances are different, provided the two sample sizes, n_1 and n_2, are equal.

PROBLEMS

1. An unbiased estimate
   A. is equal to the true parameter
   B. has the smallest variance of all possible estimates
   C. is never an efficient estimate
   D. has mean equal to the true parameter
   E. is always a maximum likelihood estimate

2. If the standard error of the mean obtained from a sample of nine observations is quoted as being three units, then nine units is
   A. the true variance of the population
   B. the estimated variance of the population
   C. the true standard deviation of the population
   D. the estimated standard deviation of the population
   E. none of the above

3. We often choose different estimators for different statistical problems. An estimator that has good properties, even when the assumptions made in choosing it over its competitors are false, is said to be
   A. unbiased
   B. efficient
   C. maximum likelihood
   D. robust
   E. minimum variance

4. For samples comprising a very large number of study units, all the following are true of maximum likelihood estimators except
   A. they are unbiased
   B. they are efficient
   C. they are normally distributed
   D. they are unique in all applications
   E. they are suitable for constructing confidence intervals

5. Parameter values that make the data we observe "most likely" to occur in a sample we have obtained are called
   A. asymptotic estimates
   B. confidence limits
   C. robust estimates
   D. maximum likelihood estimates
   E. interval estimates

6. An experimenter reports that on the basis of a sample of size 10, he calculates the 95% confidence limits for mean height to be 66 and 74 inches. Assuming his calculations are correct, this result is to be interpreted as meaning
   A. there is a 95% probability that the population mean height lies between 66 and 74 inches
   B. we have 95% confidence that a person's height lies between 66 and 74 inches
   C. we have 95% confidence that the population mean height lies between 66 and 74 inches
   D. 95% of the population has a height between 66 and 74 inches
   E. none of the above

7. A 99% confidence interval for a mean
   A. is wider than a 95% confidence interval
   B. is narrower than a 95% confidence interval
   C. includes the mean with 99% probability
   D. excludes the mean with 99% probability
   E. is obtained as the sample average plus two standard deviations

8. In a series of journal articles, investigator A reported her data, which are about normally distributed, in terms of a mean plus or minus two standard deviations, while investigator B reported his data in terms of a mean plus or minus two standard errors of the mean. The difference between the two methods is
   A. investigator A is estimating the extreme percentiles, whereas investigator B is estimating the most usual percentiles
   B. investigator A is estimating the range that she thinks contains 95% of the means, whereas investigator B is estimating the range that he thinks contains 95% of the medians
   C. investigator A is estimating the range that she thinks contains about 95% of her data values, whereas investigator B is estimating the range that he thinks (with 95% confidence) contains the true mean being estimated
   D. investigators A and B are really estimating the same range, but are just using different systems of reporting
   E. none of the above

9. A 95% confidence interval implies that
   A. the t-test gives correct intervals 95% of the time
   B. if we repeatedly select random samples and construct such interval estimates, 95 out of 100 of the intervals would be expected to bracket the true parameter
   C. the hypothesis will be false in 95 out of 100 such intervals
   D. the probability that the interval is false is 95%
   E. there is a 95% probability that the underlying distribution is normal

10. For Student's t-distribution with one degree of freedom, all the following are true except
   A. has variance 1
   B. has fatter tails than a normal distribution
   C. can be used to obtain confidence limits for the mean of a normal distribution from a sample of two observations
   D. has mean 0
   E. is symmetric

11. In a sample of 100 normal women between the ages of 25 and 29 years, systolic blood pressure was found to follow a normal distribution. If the sample mean pressure was 120 mm Hg and the standard deviation was 10 mm Hg, what interval of blood pressure would represent an approximate 95% confidence interval for the true mean?
   A. 118 to 122 mm Hg
   B. 100 to 140 mm Hg
   C. 119 to 121 mm Hg
   D. 110 to 130 mm Hg
   E. 90 to 150 mm Hg

12. An investigator is interested in the mean cholesterol level of patients with myocardial infarction. On the basis of a random sample of 50 such patients, a 95% confidence interval for the mean has a width of 10 mg/dl. How large a sample would be expected to have given an interval with a width of about 5 mg/dl?
   A. 100
   B. 200
   C. 300
   D. 400
   E. 800

13. A researcher is interested in the population variability of a normally distributed trait and finds two estimates of its standard deviation in the literature. These two estimates are similar, and the researcher wishes to average them to obtain one overall estimate. The best procedure is to

A. take the simple average of the estimated standard deviations
B. take a weighted average of the estimated standard deviations, weighting them by their degrees of freedom
C. take a simple average of the squares of the estimated standard deviations, and then take the square root of the result
D. take a weighted average of the squares of the estimated standard deviations, weighting them by their degrees of freedom, and then take the square root of the result
E. none of the above

14. A sample of five numbers is selected, and it is found that their mean is $\bar{y} = 24$. Given this information, the number of degrees of freedom available for computing the sample standard deviation is
A. 1
B. 2
C. 3
D. 4
E. 5

15. An investigator wishes to estimate the mean cholesterol level in a pediatric population. He decides, on the basis of a small sample, to calculate 95% confidence limits for the population mean. Since the data appear to be normally distributed, the appropriate statistical distribution to use in calculating the confidence interval is the
A. normal distribution
B. t-distribution
C. F-distribution
D. binomial distribution
E. Poisson distribution

CHAPTER SEVEN

Significance Tests and Tests of Hypotheses

CHAPTER SEVEN

Key Concepts

research hypothesis, null hypothesis, test criterion, significance level, p-value

one-sided (one-tail) test, two-sided (two-tail) test

F-distribution

two-sample t-test, paired, or matched pair, t-test

distribution-free methods:
- *rank sum test*
- *signed rank sum test*
- *sign test*

Type I error, validity

Type II error, power

Symbols and Abbreviations

d	difference between paired values
F	percentile of the F distribution or the corresponding test statistic
H_0	null hypothesis
p	p-value (also denoted P)
T	rank sum statistic
α	probability of type I error; significance level (Greek letter alpha)
β	probability of type II error; complement of power (Greek letter beta)

Significance Tests and Tests of Hypotheses

PRINCIPLE OF SIGNIFICANCE TESTING

A hypothesis is a contention that may or may not be true, but is provisionally assumed to be true until new evidence suggests otherwise. A hypothesis may be proposed from a hunch, from a guess, or on the basis of preliminary observations. A statistical hypothesis is a contention about a population, and we investigate it by performing a study on a sample collected from that population. We then examine the sample information to see how consistent the "data" are with the hypothesis under question; if there are discrepancies, we tend to disbelieve the hypothesis and reject it. So the question arises, How inconsistent with the hypothesis do the sample data have to be before we are prepared to reject the hypothesis? It is to answer questions such as this that we use statistical or **significance** tests. In general, three steps are taken in performing a significance test:

1. Convert the **research hypothesis** to be investigated into a specific statistical **null hypothesis.** The null hypothesis is a specific hypothesis that we try to disprove. It is usually expressed in terms of population parameters. For example, suppose that our research hypothesis is that a particular drug will lower blood pressure. We randomly assign patients to two groups: group 1 to receive the drug and group 2 to act as controls without the drug. Our research hypothesis is that after treatment, the mean blood pressure of group 1, μ_1, will be less than that of group 2, μ_2. In this situation the specific null hypothesis that we try to disprove is $\mu_1 = \mu_2$. In another situation, we might disbelieve a

claim that a new surgical procedure will cure at least 60% of cases of a particular type of cancer, and our research hypothesis would be that the probability of cure, π, is less than this. The null hypothesis that we would try to disprove is $\pi = 0.6$. Another null hypothesis might be that two variances are equal (i.e., $\sigma_1^2 = \sigma_2^2$). Notice that these null hypotheses can be expressed as a function of parameters equaling zero, and hence the terminology **null** hypothesis:

$$\mu_1 - \mu_2 = 0, \qquad \pi - 0.6 = 0, \qquad \sigma_1^2 - \sigma_2^2 = 0$$

2. Decide on an appropriate **test criterion** to be calculated from the sample values. We view this calculated quantity as one particular value of a random variable that takes on different values in different samples. Our statistical test will utilize the fact that we know how this quantity is distributed from sample to sample if the null hypothesis is true. This distribution is the sampling distribution of the test criterion under the null hypothesis and is often referred to as the null distribution. We shall give examples of several test criteria later in this chapter.

3. Calculate the test criterion from the sample and compare it with its sampling distribution to quantify how "probable" it is under the null hypothesis. We summarize the probability by a quantity known as the **p-value:** the probability of the observed or any more extreme sample occurring, if the null hypothesis is true.

PRINCIPLE OF HYPOTHESIS TESTING

If the p-value is large, we conclude that the evidence is insufficient to reject the null hypothesis; for the time being, we keep the null hypothesis. If, on the other hand, it is small, we would tend to reject the null hypothesis in favor of the research hypothesis. In significance testing we end up with a p-value, which is a measure of how unlikely it is to obtain the results we obtained if in fact the null hypothesis is true. In hypothesis testing, on the other hand, we end up either accepting or rejecting the null hypothesis outright, but with the knowledge that, if the null hypothesis is true, the probability that we reject it is no greater than a predetermined probability called the **significance level.** Significance testing and hypothesis testing are closely related and steps 1 and 2 indicated above are identical in the two procedures. The difference lies in how we interpret, and act on, the results.

Thus in hypothesis testing, instead of step 3 indicated above for significance testing, we do the following:

3. Before any data are collected, decide on a particular significance level—the probability with which we are prepared to make a wrong decision if the null hypothesis is true.

4. Calculate the test criterion from the sample and compare it with its sampling distribution. This is done in a similar manner as for significance testing, but, as we shall illustrate with some examples, we end up either accepting or rejecting the null hypothesis.

TESTING A POPULATION MEAN

As an example of a particular significance test, suppose our research hypothesis is that the mean weight of adult male patients who have been on a weight reduction program is less than 200 lb. We wish to determine whether this is so. The three steps are as follows:

1. We try to disprove that the mean weight is 200 lb. The null hypothesis we take is therefore that the mean weight is 200 lb. We can express this null hypothesis as $\mu = 200$.

2. Suppose now that, among male patients who have been on the program, weight is normally distributed with mean 200 lb. We weigh a random sample of such men and calculate the sample mean $\bar{y}$ and standard deviation s_Y. Then we know that

$$\frac{\bar{Y} - \mu}{S_{\bar{Y}}} = \frac{\bar{Y} - 200}{S_{\bar{Y}}}, \qquad \text{where } S_{\bar{Y}} = \frac{S_Y}{\sqrt{n}},$$

follows Student's t-distribution with $n - 1$ degrees of freedom. We therefore use

$$t = \frac{\bar{y} - 200}{s_{\bar{Y}}} = \frac{\bar{Y} - \mu}{s/\sqrt{n}}$$

as our test criterion: we know its distribution from sample to sample if the mean of Y is in fact 200 and Y is normally distributed.

3. Suppose our sample consisted of $n = 10$ men on the program, and we calculated from this sample $\bar{y} = 184$ and $s_Y = 26.5$ (and hence, $s_{\bar{Y}} =$

n-1 = df ?

$26.5/\sqrt{10} = 8.38$). Thus

$$t = \frac{184 - 200}{8.38} = -1.9$$

We now quantify the "probability" of such a finding as follows. From a table of Student's t-distribution (see Table 2, Appendix 2) we find, for $10 - 1 = 9$ degrees of freedom, the following percentiles:

%:	2.5	5	95	97.5
t-value:	−2.262	−1.833	1.833	2.262

(The table gives us $t_{95} = -t_5 = 1.833$ and $t_{97.5} = -t_{2.5} = 2.262$ where t_q is the qth percentile of the t-distribution.) Thus the value we found, −1.9, lies between the 2.5th and 5th percentiles. Pictorially, the situation is as illustrated in Figure 7–1. If the value of t had been somewhere near 0, this would be a "probable" value and there would be no reason to suspect the null hypothesis. The t-value of −1.9 lies, however, below the 5th percentile (i.e., if the null hypothesis is true, the probability is less than 0.05 that t should be as far to the left as −1.9). We are therefore faced with what the famous statistician Sir Ronald Fisher (1890–1962) called a "logical disjunction": either the null hypothesis is not true, or, if it is true, we have observed a rare event—one that has less than 5% probability of occurring simply by chance. Symbolically this is often written, with no further explanation, $p < 0.05$. It is understood that p (also often denoted P) stands for the probability of observing what we actually did observe, or anything more extreme, if the null hypothesis is true. Thus, in our example, p is the area to the left of −1.9 under the curve of a t-distribution with

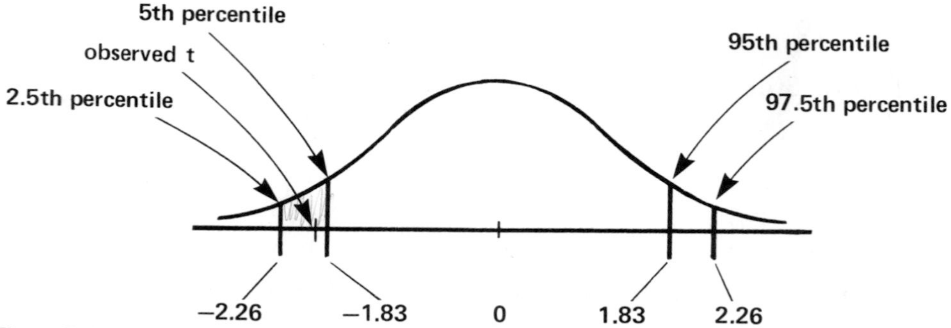

Figure 7–1. Comparison of the observed t = −1.9 with Student's t-distribution with nine degrees of freedom.

9 degrees of freedom. The area is about 4.5%, and this fact can be expressed as $p \doteq 0.045$.

Because the p-value is no larger than 0.05, we say that the result is *significant at the 5% level,* or that the mean is significantly less than 200 lb at the 5% level. We can also say the result is significant at the 4.5% level, because the p-value is no larger than 4.5%. Similarly, we can say that the result is significant at the 0.1 level, because 0.1 is larger than p. In fact we can say that the result is significant at any level greater than 4.5%, and not significant at any level less than 4.5%.

Notice, however, what the null hypothesis was (i.e., what was assumed to obtain the distribution of the test criterion). We assumed the theoretical population of weights had both a normal distribution and a mean of 200 lb. We also assumed, in order to arrive at a statistic that should theoretically follow the t-distribution, that the n weights available constitute a random sample from that population. All these assumptions are part of the null hypothesis that is being tested, and departure from any one of them could be the cause of a significant result. Provided we do have a random sample and a normal distribution, however, either we have observed an unlikely outcome ($p = 0.045$) or, contrary to our initial assumption, the mean is less than 200 lb.

Rather than perform a significance test, the result of which is a p-value, many investigators perform a test of hypothesis: at the beginning of a study, before any data are collected, they pick a specific level (often 5% or 1%) as a cutoff, and decide to "reject" the null hypothesis for any result significant at that level. It is a common (but arbitrary) convention to consider any value of p greater than 0.05 as "not significant." The idea behind this is that one should not place too much faith in a result that, by chance alone, would be expected to occur with a probability greater than 1 in 20. Other conventional phrases that are sometimes used are

$0.01 < p < 0.05$: "significant"
$0.001 < p < 0.01$: "highly significant"
$p < 0.001$: "very highly significant"

This convention is quite arbitrary, and arose originally because the cumulative probabilities of the various sampling distributions (such as the t-distribution) are not easy to calculate, and so had to be tabulated. Typical tables, so as not to be too bulky, include just a few percentiles, such as the 90th, 95th, 97.5th, 99th, and 99.5th percentiles, corresponding to the tail probabilities 0.1, 0.05, 0.025, 0.01, and 0.005 for one tail of the distribution. Now that computers and calculators are commonplace, however, it is becoming more and more common to calculate and quote the actual value of p. Although many investigators still use a significance level of 0.05 for testing hypotheses, it is clearly absurd to quote a result for which p = 0.049 as "significant" and one for which p = 0.051 merely as "not significant": it is far more informative to quote the actual p-values, which an intelligent reader can see are virtually identical in this case.

Note that we have defined the meaning of "significant" in terms of probability: in this sense it is a technical term, always to be interpreted in this precise way. This is often emphasized by saying, for example, "the result was statistically significant." Such a phrase, however, although making it clear that significance in a probability sense is meant, is completely meaningless unless the level of significance is also given. (The result of every experiment is statistically significant at the 100% level, because the significance level can be any probability larger than p!) It is important to realize that statistical significance is far different from biological significance. If we examine a large enough sample, even a biologically trivial difference can be made to be statistically significant. Conversely, a difference that is large enough to be of great biological significance can be statistically "not significant" if a very small sample size is used. We shall come back to this point at the end of the chapter.

Notice carefully the definition of p: the probability of observing what we actually did observe, *or anything more extreme*, if the null hypothesis is true. By "anything more extreme," we mean any result that would alert us even more (than the result we actually observed) to the possibility that our research hypothesis, and not the null hypothesis, is true. In our example, the research hypothesis is that the mean weight is less than 200 lb; therefore a sample mean less than 200 lb (which would result in a negative value of t) could suggest that the research hypothesis is true, and any value of t less than (i.e., more negative than) −1.9 would alert us even more to the possibility that the null hypothesis is not true. A t-value of +2.5, on the other hand, would certainly not suggest that the research hypothesis is true.

ONE-SIDED VERSUS TWO-SIDED TESTS

Now suppose, in the above example, we had wished to determine whether the mean weight is *different from* 200 lb, rather than *is less than*

200 lb. Expressed in this way, our research hypothesis is that there is a difference, but in an unspecified direction. We believe that the program will affect weight but are unwilling to state ahead of time whether the final weight will be more or less than 200 lb. Any extreme deviation from 200 lb, whether positive or negative, would suggest that the null hypothesis is not true. Had this been the case, not only would a t-value less than -1.9 be more extreme, but so also would any t-value *greater than* $+1.9$. Thus, because of the symmetry of the t-distribution, the value of p would be *double* 4.5%, that is, 9%: we add up the probability to the left of -1.9 and the probability to the right of $+1.9$ (i.e., the probabilities in both tails of the distribution).

Thus the significance level depends on what we had in mind before we actually sampled the population. If we knew beforehand that the weight-reduction program could not lead to a weight above 200 lb, our question would be whether the mean weight is *less* than 200 lb. We would perform what is known as a **one-sided, or one-tail, test,** using only the left-hand tail of the t-distribution; and we would report the resulting $t = -1.9$ as being significant at the 5% level. If, on the other hand, we had no idea originally whether the program would lead to a mean weight above or below 200 lb, the question of interest would be whether or not the true mean is *different from* 200 lb. We would then perform a **two-sided, or two-tail, test,** using the probabilities in both the left- and right-hand tails of the t-distribution; and for our example, a t-value of -1.9 would then not be significant at the 5% level, although it would be significant at the 10% level ($p = 0.09$).

There is a close connection between a two-sided test and a confidence interval. Let us calculate the 95% and 90% confidence intervals for the mean weight of men on the weight-reduction program. We have

$$n = 10 \qquad \bar{y} = 184 \qquad s_{\bar{Y}} = 8.38$$

From Table 2, Appendix 2, we see that for 9 degrees of freedom, $t_{97.5} = 2.262$ and $t_{95} = 1.833$. We therefore have

95% confidence interval: $184 \pm (2.262)8.38$, or 165.0 to 203.0
90% confidence interval: $184 \pm (1.833)8.38$, or 168.6 to 199.4.

The 95% interval includes the value 200, whereas the 90% interval does not. In general, a sample estimate (184 in this example) will be significantly different from a hypothesized value (200) if and only if the corresponding confidence interval for the parameter does *not* include that value. A 95% confidence interval corresponds to a two-sided test at the 5% significance level: the interval contains 200, and the test is not significant at the 5%

level. A 90% confidence interval corresponds to a test at the 10% significance level: the interval does not include 200, and the test is significant at the 10% level. In general a $100(1 - \alpha)$ % confidence interval corresponds to a two-sided test at the 100α % significance level.

TESTING A PROPORTION

Suppose an investigator disputes a claim that, using a new surgical procedure for a risky operation, the proportion of successes is at least 0.6. The investigator is interested in showing that the proportion is in fact less than 0.6. The three steps for a significance test of this research hypothesis are then as follows:

1. The null hypothesis is taken to be that the proportion of successes is 0.6 (i.e., $\pi = 0.6$); we shall see whether the data are consistent with this null hypothesis, or whether we should reject it in favor of $\pi < 0.6$.
2. Suppose Y, the number of successes, is binomially distributed. Then we can use Y as our test criterion; once the sample size is determined, its distribution is known if in fact $\pi = 0.6$ and Y is binomially distributed.
3. We select a random sample of operations in which this new procedure is used, say $n = 10$ operations, and find, let us suppose, $y = 3$ successes. From the binomial distribution with $n = 10$ and $\pi = 0.6$, the probability of each possible number of successes is as follows:

Number of Successes	Probability
0	0.0001
1	0.0016
2	0.0106
3	0.0425
4	0.1114
5	0.2007
6	0.2508
7	0.2150
8	0.1209
9	0.0404
10	0.0060
Total	1.0000

To determine the p-value, we sum the probabilities of all outcomes as extreme or more extreme than the one observed. The "extreme outcomes" are those that suggest the research hypothesis is true and alert us, even more than the sample itself, to the possibility that the null hypothesis is false. If $\pi = 0.6$, we expect, on an average, 6 successes in 10 operations. A series of 10 operations with fewer successes would suggest that $\pi < 0.6$, and hence that the research hypothesis is true and the null hypothesis is false. Thus, 0, 1, 2, or 3 successes would be as extreme or more extreme than the 3 observed. We sum the probabilities of these four outcomes to obtain $0.0001 + 0.0016 + 0.0106 + 0.0425 = 0.0548$ (i.e., $p = 0.0548$).

We find it difficult to believe that we would be so unlucky as to obtain an outcome as rare as this if π is 0.6, as claimed. We believe, rather, that $\pi < 0.6$, since this would give rise to larger probabilities of observing 0, 1, 2, or 3 successes in 10 operations. We are therefore inclined to reject the null hypothesis and conclude that the probability of success using the new procedure is less than 0.6. Specifically, we can say that the observed proportion of successes, 3 out of 10, or 0.3, is significantly less than 0.6 at the 6% level.

Let us suppose, for illustrative purposes, that the sample of 10 operations had resulted in $y = 8$ successes. Such an outcome would be consistent with the null hypothesis. All y values less than 8 would be closer to the research hypothesis than the null hypothesis, and so the p-value for such an outcome would be

$$\begin{aligned} P(0) + P(1) + \cdots + P(8) &= 1 - P(9) - P(10) \\ &= 1 - 0.0404 - 0.0060 \\ &= 0.9536 \end{aligned}$$

In this instance it is obvious that we should retain the null hypothesis (i.e., the data are consistent with the hypothesis that the probability of a success using the new procedure is at least 0.6). But note carefully that "being consistent with" a hypothesis is not the same as "is strong evidence for" a hypothesis. We would be much more convinced that the hypothesis is true if there had been 800 successes out of 1,000 operations. "Retaining" or "accepting" the null hypothesis merely means that we have insufficient evidence to reject it—not that it is true.

If the number of operations had been large, much effort would be needed to calculate, from the formula for the binomial distribution, the probability of each possible more extreme outcome. Suppose, for example, there had been $n = 100$ operations and the number of successes was $y = 30$, so that the proportion of successes in the sample is still 0.3, as before. In this case, it would be necessary to calculate the probabilities of 0, 1, 2, . . . right on up to 30 successes, in order to obtain the exact p-value. But in such a case we can take advantage of the fact that n is large. When n is large we know that the average number of successes per operation, Y/n (i.e., the proportion of successes) is approximately normally distributed. The three steps for a test are then as follows:

1. The null hypothesis is $\pi = 0.6$, as before.

2. Since both $n\pi$ and $n(1 - \pi)$ are greater than 5 under the null hypothesis (they are 60 and 40, respectively), we assume that Y/n is normally distributed with mean 0.6 and standard deviation $\sqrt{0.6(1 - 0.6)/n} = \sqrt{0.24/100} = 0.049$. Thus, under the null hypothesis, the standardized variable

$$Z = \frac{Y/n - 0.6}{0.049}$$

approximately follows a standard normal distribution and can be used as the test criterion.

3. We observe $y/n = 0.3$ and hence

$$z = \frac{0.3 - 0.6}{0.049} = -6.12$$

and any value of z less than this is more extreme (i.e., even less consistent with the null hypothesis). Consulting the probabilities for the standard normal distribution (first line of Table 1, Appendix 2), we find

$$P(Z < -3.49) = 0.0002$$

and $P(Z < -6.12)$ must be even smaller than this. We are thus led to reject the null hypothesis at an even smaller significance level.

We can see the improvement in the normal approximation of the binomial distribution with increasing sample sizes from the following probabilities, calculated on the assumption that $\pi = 0.6$:

Sample Size	$P(Y/n \leq 0.3)$	
	Binomial	Normal
10	0.0548	0.0262
20	0.0065	0.0031
30	0.0002	0.0004

If we restrict our attention to the first two decimal places, the difference in p-values is about 0.03 for a sample of size 10, but less than 0.01 for a sample of size 20 or larger.

The binomial distribution, or the normal approximation in the case of a large sample, can be used in a similar manner to test hypotheses about any percentile of a population distribution. As an example, suppose we wish to test the hypothesis that the median (i.e., the 50th percentile) of a population distribution is equal to a particular hypothetical value. A random sample of n observations from the distribution can be classified into two groups: those above the hypothetical median and those below the hypothetical median. We then simply test the null hypothesis that the proportion above the median (or equivalently, the proportion below the median) is equal to 0.5 (i.e., $\pi = 0.5$). This is simply a special case of testing a hypothesis about a proportion in a population. If we ask whether the population median is *smaller* than the hypothesized value, we perform a one-sided test similar to the one performed above. If we ask whether it is *larger*, we similarly perform a one-sided test, but the appropriate p-value is obtained by summing the probabilities in the other tail. If, finally, we ask whether the median is *different* from the hypothesized value, a *two-sided* test is performed, summing the probabilities of the extreme outcomes in both tails to determine whether to reject the null hypothesis.

TESTING THE EQUALITY OF TWO VARIANCES

Often, we wish to compare two samples. We may ask, for example, whether the distribution of serum cholesterol levels is the same for males and females in a set of patients. First, we could ask whether the distribution in each population is normal, and there are various tests for this. If we find the assumption of normality reasonable, we might then assume normality and ask whether the variance is the same in both populations from which the samples come. Let the two sample variances be s_1^2 and s_2^2. Then an appropriate criterion to test the null hypothesis that the two population variances are equal is the ratio s_1^2/s_2^2. Provided the distribution in each population is normal, under the null hypothesis this statistic comes from an **F-distribution,** named in honor of Sir Ronald A. Fisher. The F-distribution is a two-parameter distribution, the two parameters being the number of degrees of freedom in the numerator (s_1^2) and the number of degrees of freedom in the denominator (s_2^2). If the sample sizes of the two groups are n_1 and n_2, then the numbers of degrees of freedom are, respectively, $n_1 - 1$ and $n_2 - 1$. A few percentile points of the F-distribution are given in Table 3 of Appendix 2. The number of degrees of freedom along the *top* of the table corresponds to that in the *top* of the F-ratio ($n_1 - 1$), whereas that along the side of the table corresponds to that in the bottom of the F-ratio. (All tables of the F-distribution follow this convention.) The table is appropriate for testing the null hypothesis $\sigma_1^2 = \sigma_2^2$ against the alternative $\sigma_1^2 > \sigma_2^2$: large values of F are significant. This is a one-sided test. If we wish to perform a two-sided test, we put the larger of the two sample variances, s_1^2 or s_2^2, on top and *double* the tail probability indicated by the table.

A numerical example will illustrate the procedure. Suppose we have a sample of $n_1 = 10$ men and $n_2 = 25$ women, with sample variances $s_1^2 = 30.3$ and $s_2^2 = 69.7$, respectively, for a trait of interest. We wish to test the null hypothesis that the two population variances are equal (i.e., $\sigma_1^2 = \sigma_2^2$). We have no prior knowledge to suggest which might be larger, and so we wish to perform a two-sided test. We therefore put the larger sample variance on top to calculate the ratio

$$\frac{69.7}{30.3} = 2.30$$

There are $25 - 1 = 24$ degrees of freedom in the top of this ratio and $10 - 1 = 9$ degrees of freedom in the bottom. Looking at the columns headed 24, and the rows labeled 9, in the table of the F distribution (Table 3, Appendix 2), we find the following percentiles:

%:	90	95	97.5	99	99.5
F-value:	2.28	2.90	3.61	4.73	5.73

The observed ratio, 2.30, lies between the 90th and 95th percentiles, corresponding to tail probabilities of 0.1 and 0.05. Because we wish to perform a two-sided test, we double these probabilities to obtain the p-value. The result would thus be quoted as $0.1 < p < 0.2$, or as $p \doteq 0.2$ (since 2.30 is close to 2.28).

In this instance we might decide it is reasonable to assume that the two variances are equal. As we learned in Chapter 6, the common, or "pooled," variance is then estimated as

$$s_p^2 = \frac{(n_1 - 1)s_1^2 + (n_2 - 1)s_2^2}{n_1 + n_2 - 2}$$

which in this case is

$$\frac{(9)30.3 + (24)69.7}{9 + 24} = 59.0$$

This estimate is unbiased.

Note once again that the null hypothesis is not simply that the two variances are equal, although the F-test is often described as a test for the equality of two variances. For the test criterion to follow an F-distribution, each sample must also be made up of normally distributed random variables. In other words, the null hypothesis is that the two samples are made up of independent observations from two normally distributed populations with the same variance. The distribution of the F-statistic is known to be especially sensitive to nonnormality, so a significant result could be due to nonnormality and have nothing to do with whether or not the population variances are equal.

TESTING THE EQUALITY OF TWO MEANS

Suppose now we can assume that the random variables of interest are normally distributed, with the same variance in the two populations. Then we can use a **two-sample t-test** to test whether the means of the random variable are significantly different in the two populations. Let $\overline{Y}_1$ and $\overline{Y}_2$ be the two sample means. Then, under the null hypothesis that the two population means are the same, $\overline{Y}_1 - \overline{Y}_2$ will be normally distributed with mean zero. Furthermore, provided the observations in the two groups are independent (taking separate random samples from the two populations will ensure this), the variance of $\overline{Y}_1 - \overline{Y}_2$ will be $\sigma^2_{\overline{Y}_1} + \sigma^2_{\overline{Y}_2}$ (i.e., $\sigma^2/n_1 + \sigma^2/n_2$) where σ^2 is the common variance. Thus

$$\frac{\overline{Y}_1 - \overline{Y}_2}{\sqrt{\sigma^2_{\overline{Y}_1} + \sigma^2_{\overline{Y}_2}}} = \frac{\overline{Y}_1 - \overline{Y}_2}{\sigma\sqrt{\frac{1}{n_1} + \frac{1}{n_2}}}$$

will follow a standard normal distribution, and analogously,

$$\frac{\overline{Y}_1 - \overline{Y}_2}{S_p\sqrt{\frac{1}{n_1} + \frac{1}{n_2}}}$$

will follow a t-distribution with $n_1 + n_2 - 2$ degrees of freedom; in this formula we have replaced σ by S_p, the square root of the pooled variance estimator. Thus we calculate

$$t = \frac{\overline{y}_1 - \overline{y}_2}{s_p\sqrt{\frac{1}{n_1} + \frac{1}{n_2}}}$$

and compare it with percentiles of the t-distribution with $n_1 + n_2 - 2$ degrees of freedom. As before, if $n_1 + n_2 - 2$ is greater than 30, the percentiles are virtually the same as for the standard normal distribution.

Suppose, for our example of $n_1 = 10$ men and $n_2 = 25$ women, we found the sample means $\overline{y}_1 = 101.05$ and $\overline{y}_2 = 95.20$. We have already seen that $s_p^2 = 59.0$, and so $s_p = \sqrt{59.0} = 7.68$. To test whether the means are significantly different, we calculate

$$t = \frac{\overline{y}_1 - \overline{y}_2}{s_p\sqrt{\frac{1}{n_1} + \frac{1}{n_2}}} = \frac{101.05 - 95.20}{7.68\sqrt{\frac{1}{10} + \frac{1}{25}}} = 2.04$$

There are $9 + 24 = 33$ degrees of freedom, so 2.04 is approximately the 97.5th percentile of the t-distribution. Thus, $p = 0.025$ for a one-sided test and $p = 0.05$ for a two-sided test.

Note carefully the assumption that the two samples are independent. Often this assumption is purposely violated in designing an experiment to compare two groups. Cholesterol levels, for example, change with age; so if our sample of men were very different in age from our sample of women, we would not know whether any difference that we found was due to gender or to age (i.e., these two effects would be confounded). To obviate this, we could take a sample of men and women who are individually matched for age. We would still have two samples, n men and n women, but they would no longer be independent. We would expect the pairs of cholesterol levels to be correlated, in the sense that the cholesterol levels of a man and woman who are the same age will tend to be more alike than those of a man and woman who are different ages (the term "correlation" will be defined more precisely in Chapter 9). In the case where individuals are matched, an appropriate test for a mean difference between the two populations would be the **paired t-test,** or **matched pair t-test.** We would pair the men and women and find the difference in cholesterol level for each pair (taking care always to subtract the cholesterol level of a person of the same sex—so some of the differences may be negative). Let the difference be d. Then we have n d's, and, if the null hypothesis of no mean difference between male and female is true, the d's are expected to have mean zero. Thus we would calculate

$$t = \frac{\bar{d}}{s_{\bar{D}}} = \frac{\bar{d}}{s_{\bar{D}}/\sqrt{n}}$$

where $\bar{d}$ is the mean of the n d's and s_D their estimated standard deviation, and compare this with percentiles of the t-distribution with $n - 1$ degrees of freedom. This test assumes that the differences (the d's) are normally distributed. Notice our continued use of capital letters to denote random variables and lower case letters for their specific values. Thus, D is the random variable denoting a difference and s_D denotes the estimated standard deviation of D.

TESTING THE EQUALITY OF TWO MEDIANS

The median of a normal distribution is the same as its mean. It follows that if our sample data come from normal distributions, testing for the equality of two medians is the same as testing for the equality of two means. If our samples do not come from normal distributions, however, we should not use the t-distribution as indicated above to test for the equality of two means in small samples. Furthermore, if the population distributions are at all skewed, the medians are better parameters of central tendency. We should then probably be more interested in testing the equality of the two medians than in testing the equality of the two population means. In this section we shall outline methods of doing this without making distributional assumptions such as normality. For this reason the methods we shall describe are sometimes called **distribution-free methods.** We shall indicate statistics that can be used as criteria for the tests and note that for large samples they are approximately normally distributed, regardless of the distributions of the underlying populations. It is beyond the scope of this book to discuss the distribution of all these statistics in small samples, but you should be aware that appropriate tables are available for such situations.

First, suppose we have two independent samples: n_1 observations from one population and n_2 observations from a second population. Wilcoxon's **rank sum test** is the appropriate test in this situation, provided we can assume that the distributions in the two populations, while perhaps having different medians, have the same (arbitrary) shape. The observations in the two samples are first considered as a single set of $n_1 + n_2$ numbers, and arranged in order from the smallest to largest. Each observation is assigned a rank: 1 for the smallest observation, 2 for the next smallest, and so on, until $n_1 + n_2$ is assigned to the largest observation. The ranks of the observations in the smaller of the two samples are then summed, and this is the statistic, which we denote T, whose distribution is known under the null hypothesis. Percentile points of the distribution of T have been tabulated, but for large samples we can assume that T is approximately normally distributed. (An alternative method of calculating a test criterion in this situation is called the Mann-Whitney test. Wilcoxon's test and the Mann-Whitney test are equivalent and so we omit describing the calculation of the latter. It is also of interest to note that these two tests are equivalent, in large samples, to performing a two-sample t-test on the ranks of the observations).

As an example, suppose we wish to compare the median serum cholesterol levels in milligrams per deciliter for two groups of students, based on the following samples:

Sample 1, $n_1 = 6$: 58, 92, 47, 126, 53, 85

Sample 2, $n_2 = 7$: 87, 199, 124, 83, 115, 68, 156

The combined set of numbers and their corresponding ranks are

47	53	58	68	83	85	87	92	115	124	126	156	199
<u>1</u>	<u>2</u>	<u>3</u>	4	5	<u>6</u>	7	<u>8</u>	9	10	<u>11</u>	12	13

The underlined ranks correspond to the smaller sample, their sum being 1 + 2 + 3 + 6 + 8 + 11 = 31. Although the samples are not really large enough to justify using the large sample normal approximation, we shall nevertheless use these data to illustrate the method. We standardize T by subtracting its mean and dividing by its standard deviation, these being derived under the null hypothesis that the two medians are equal. The result is then compared with percentiles of the normal distribution. If $n_1 \leq n_2$ (i.e., n_1 is the size of the smaller sample) it is shown in Appendix 1 that the mean value of T is $n_1(n_1 + n_2 + 1)/2$, which in this case is $6(6 + 7 + 1)/2 = 42$. Also, it can be shown that the standard deviation of T is $\sqrt{n_1 n_2 (n_1 + n_2 + 1)/12}$, which in our example is $\sqrt{(6)(7)(14)/12} = 7$. Thus we calculate the standardized criterion

$$z = \frac{31 - 42}{7} = -1.57$$

Looking this up in the table (Table 1, Appendix 2), we find it lies at the 5.82 percentile, which for a two-sided test corresponds to $p = 0.1164$. In fact, tables that give the percentiles of the exact distribution of T also indicate that $0.1 < p < 0.2$, so in this instance the normal approximation does not mislead us.

Let us now suppose the samples were taken in such a way that the data are paired, with each pair consisting of one observation from each population. The study units might be paired, for example, in a randomized-blocks experimental design in which each block consists of only two subjects (of the same age and sex) randomly assigned to one or the other of two treatments. Paired observations could also arise in situations in which the same subject is measured before and after treatment.

Since the study units are paired, the difference in the observation of interest can be computed for each pair, taking care always to calculate the difference in the same direction. These differences can then be analyzed by Wilcoxon's **signed rank (sum) test** as follows: First we rank the differences from smallest to largest, without regard to the sign of the difference. Then we sum the ranks of the positive and negative differences separately, and the smaller of these two numbers is entered into an appropriate table to determine the p-value. For large samples we can again use a normal approximation, using the fact that under the null hypothesis the mean and the standard deviation of the sum depend only on the number of pairs n. As an example of this test, let us suppose that eight identical twin pairs were studied to investigate the effect of a triglyceride-lowering drug. A member of each pair was randomly assigned to either the active drug or a placebo, with the other member of the pair receiving the other treatment. The resulting data are as follows (triglyceride values are in mg/dl):

Twin pair	1	2	3	4	5	6	7	8
Placebo twin	71	65	126	111	249	198	57	97
Active drug twin	69	52	129	75	226	181	46	93
Difference	2	13	−3	36	23	17	11	4
Rank (ignoring sign)	1	5	2	8	7	6	4	3

The sum of the ranks of the positive differences is $1 + 5 + 8 + 7 + 6 + 4 + 3 = 34$, and that of the negative differences (there is only one) is 2. If we were to look up 2 in the appropriate table for $n = 8$, we would find, for a two-tail test, $0.02 < p < 0.05$. Hence, we would reject the hypothesis of equal medians at $p = 0.05$, and conclude that the active drug causes a (statistically) significant reduction in the median triglyceride level.

The large sample approximation will now be computed for this example, to illustrate the method. Under the null hypothesis, the mean of the sum is $n(n + 1)/4$ and the standard deviation is $\sqrt{n(n + 1)(2n + 1)/24}$. Thus, when $n = 8$, the mean is $8(9)/4 = 18$ and the standard deviation is $\sqrt{8(9)(17)/24} = \sqrt{51} = 7.14$. We therefore calculate

$$z = \frac{2 - 18}{7.14} = -2.24$$

which, from the normal table (Table 1, Appendix 2), lies at the 1.25 percentile. For a two-sided test, this corresponds to $p = 0.025$. Thus, even for as small a sample as this, we once again find that the normal approximation is adequate.

Let us now briefly consider another way of testing the same hypothesis. If the medians are equal in the two populations, then on an average, the number of positive differences in the sample will be the same as the number of negative differences. In other words, the mean proportion of positive differences would be 0.5 in the population (of all possible differences). Thus, we can test the null hypothesis that the proportion of positive differences, π, is 0.5. This is called the **sign test.** For a sample size $n = 8$ as in the above data, we have the following binomial distribution under the null hypothesis ($\pi = 0.5$):

Number of Minus Signs	Probability
0	0.0039
1	0.0313
2	0.1093
3	0.2188
4	0.2734
5	0.2188
6	0.1093
7	0.0313
8	0.0039

Thus, the probability of observing a result as extreme or more extreme than a single minus sign under the null hypothesis is

$$P(0) + P(1) + P(7) + P(8) = 0.0039 + 0.0313 + 0.0313 + 0.0039 = 0.0704$$

(We sum the probabilities in both tails, for a two-sided test). This result, unlike the previous one based on the same data, is no longer significant at the 5% significance level. Which result is correct, this one or the previous one? Can both be correct? To understand the difference we need to learn about how we judge different hypothesis-testing procedures.

VALIDITY AND POWER

Sometimes we have to make a definite decision one way or another about a particular hypothesis; in this situation a test of hypothesis is appropriate. Although in science we never accept a hypothesis outright, but rather continually modify our ideas and laws as new knowledge is obtained, in clinical practice we cannot afford this luxury. To understand the concepts of validity and power, it will be helpful if we consider the case in which a decision must be made, one way or the other, with the result that some wrong decisions will inevitably be made. Clearly we wish to act in such a way that the probability of making a wrong decision is minimized.

Let us suppose we perform a test of hypothesis, with the result that we either accept or reject the null hypothesis, for which from now on we shall use the abbreviation H_0. Since in the "true state of nature" H_0 must be actually true or false, we have just four possibilities, which we can depict as the entries in a 2 × 2 table as follows:

		True state of nature	
		H_0 is true	H_0 is false
Decision made	Accept H_0	O.K.	Type II error
	Reject H_0	Type I error	O.K.

In the case of two of the possibilities, denoted by the entries "**O.K.**," the decision is correct, and hence no error is made. In the case of the other two possibilities, a wrong decision, and hence an error, is made. The error may be one of two types:

Type I Rejection of the null hypothesis when in fact it is true. The probability of this happening is often denoted α [i.e., α = P(reject $H_0 | H_0$ is true)].

Type II Acceptance of the null hypothesis when in fact it is false. The probability of this happening is often denoted β [i.e., β = P(accept $H_0 | H_0$ is false)].

When performing a test of hypothesis, the significance level α is the probability of making a type I error, and we control it so that it is kept reasonably small. Suppose, for example, we decide to fix α at the value 0.05. It is our intention to accept the null hypothesis if the result is not significant at the 5% level and to reject it if it is. Then, provided our test does in fact reject H_0 in 5% of the situations in which H_0 is true, it is a **valid** test at the 5% level. A valid test is one that rejects H_0 in a proportion α of the situations in which H_0 is true, where α is the stated significance level. Suppose we have a sample of paired data from two populations that are normally distributed with the same variance. In order to test whether the two population medians are equal, we could use (1) the paired t-test, (2) the signed rank sum test, or (3) the sign test. The fact that we have normal distributions does not in any way invalidate the signed rank sum and the sign tests. Provided we use the appropriate percentiles of our test criteria (e.g., the 5th percentile or the 95th percentile for a one-sided test) to determine whether to reject the null hypothesis, we shall find that, *when it is true*, we reject H_0 with 5% probability. This will be true of all three tests; they are all *valid* tests in this situation.

Although they are all valid, the three tests nevertheless differ in the value of β, or probability of type II error. In other words, they differ in the probability of accepting the null hypothesis *when it is false* (i.e., when the medians are in fact different). In this situation we are most likely to reject the null hypothesis when using the t-test, less likely to do so when using the signed rank sum test, and least likely to do so when using the sign test. We say that the t-test is more *powerful* than the signed rank sum test, and the signed rank sum test is more powerful than the sign test. **Power** is defined as $1 - \beta$. It is the probability of rejecting the null hypothesis when in fact it is false. Note that if we identify the null hypothesis with absence of a disease, there is an analogy between the power of a statistical test and the sensitivity of a diagnostic test (defined in Chapter 3).

Now suppose we do not have normal distributions, but we can assume that the shape of the distribution is the same in both populations. If this is the case, the paired t-test may no longer be valid, and then the fact that it might be more powerful is irrelevant. Now in large samples, the t-test is fairly robust against nonnormality (i.e., the test is approximately valid even when we do not have underlying normal distributions). But this is not necessarily the case in small samples. We should not use the t-test for small samples if there is any serious doubt about the underlying population distributions being approximately normal. Note that if the samples are large enough for the t-test to be robust, then we do not need to refer the test statistic to the t distribution. We saw in the last chapter that the t-distribution with more than 30 degrees of freedom has percentiles that are about the same as those of the standard normal distribution.

If we cannot assume that the shape of the distribution is about the same in both populations, then both the paired t-test and the signed rank-sum test may be invalid, and we should use the sign test even though it is the least powerful. This illustrates a general principle of all statistical tests: The more we can assume, the more powerful our test can be. This same principle is at work in the distinction between one-sided and two-sided tests. If we are prepared to assume, prior to any experimentation, that the median of population 1 cannot be smaller than that of population 2, we can perform a one-sided test. Then, to attain a p-value less than a prespecified amount, our test criterion need not be as extreme as would be necessary for the analogous two-sided test. Thus a one-sided test is always more powerful than the corresponding two-sided test.

We often do not control β, the probability of making an error if H_0 is false, mainly because there are many ways in which H_0 can be false. It makes sense, however, to have some idea of the magnitude of this error before going to the expense of conducting an experiment. This is done by calculating $1 - \beta$, the *power* of the test, and plotting it against the "true state of nature." Just as the sensitivity of a diagnostic test to detect a disease usually increases with the severity of the disease, so the power of a statistical test usually increases with departure from H_0. For example, for the two-sided t-test of the null hypothesis $\mu_1 = \mu_2$, we can plot power against $\mu_1 - \mu_2$, as in Figure 7–2. Note that the "power curve" is symmetrical about $\mu_1 - \mu_2 = 0$ (i.e., about $\mu_1 = \mu_2$), since we are considering a two-sided test. Note also that the probability of rejecting H_0 is a minimum when H_0 is true (i.e., when $\mu_1 - \mu_2 = 0$), and that at this point it is equal to α, the significance level. The power increases as the absolute difference between μ_1 and μ_2 increases (i.e., as the departure from H_0 increases).

As you might expect, power also depends on the sample size. We can always make the probability of rejecting H_0 small by studying a small sample. Hence, not finding a significant difference, or "accepting H_0" must never be equated with the belief that H_0 is true: it merely indicates that there is insufficient evidence to reject H_0 (which may be due to the fact that H_0 is true, *or* may be due to a small sample size). It is possible to determine from the power curve how large the difference $\mu_1 - \mu_2$ must be in order for there to be a good chance of rejecting H_0 (i.e., of observing a difference that is statistically significant). Also, we could decide on a magnitude for the real difference that we should like to detect, and then plot against sample size the power of the test to detect a difference of that magnitude. This is often done before conducting a study, in order to choose an

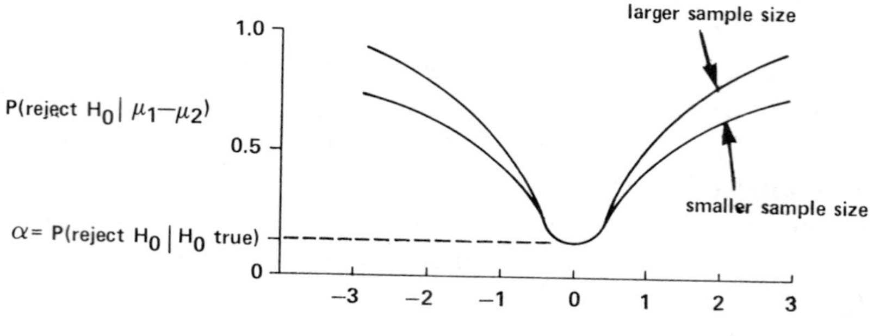

Figure 7–2. Examples of the power of the two-sided t-test for the difference between two means, μ_1 and μ_2, plotted against $\mu_1 - \mu_2$.

appropriate sample size. Power also depends on the variability of our measurements, however; the more variable they are, the less the power. For this reason power is often expressed as a function of the standardized difference $(\mu_1 - \mu_2)/\sigma$, where it is assumed that the two populations have the same standard deviation. For example, a small difference is often considered to be less than 0.2σ, a medium difference between 0.2σ and 0.8σ, and a large difference one that is larger than 0.8σ.

In summary, there are six ways to increase power when testing a hypothesis statistically:

1. Use a larger significance level. This is often less desirable since it implies a larger probability of type I error.

2. Use a larger sample size. This is more expensive.

3. Consider only larger deviations from H_0. This may be less desirable, but note that there is no point in considering differences that are too small to be biologically or medically significant.

4. Reduce variability, either by making more precise measurements or by choosing more homogeneous study units.

5. Make as many valid assumptions as possible (e.g., a one-sided test is more powerful than a two-sided test).

6. Use the most powerful test that the appropriate assumptions will allow. The most powerful test may sometimes be more expensive to compute, but it is usually cheapest in the long run.

It is also possible to increase power by using an invalid test, but this is never legitimate!

Finally, remember that if a statistical test shows that a sample difference is not significant, this does not prove that a population difference does not exist, or even that any real difference is probably small. Only the *power* of the test tells us anything about the probability of rejecting any hypothesis other than the null hypothesis. Whenever we accept the null hypothesis, a careful analysis of the power is essential. Furthermore, neither the p-value nor the power can tell us the probability that the research hypothesis is true. Equating this to the probability that H_0 is false, we can determine, using Bayes' theorem, that

$$P(H_0 \text{ is false} \mid \text{reject } H_0) = \frac{P(H_0 \text{ false})P(\text{reject } H_0 \mid H_0 \text{ false})}{P(H_0 \text{ false})P(\text{reject } H_0 \mid H_0 \text{ false}) + P(H_0 \text{ true})P(\text{reject } H_0 \mid H_0 \text{ true})}$$

in which we can substitute

$$P(\text{reject } H_0 \mid H_0 \text{ false}) = \text{power}$$

and

$$P(\text{reject } H_0 \mid H_0 \text{ true}) = \text{significance level}$$

Thus, to answer the question that is often of most interest (i.e., now that I have obtained a result that is significant at a particular level, what is the probability that H_0 is false?), we need to know, before seeing the results of the study, the prior probabilities P(H_0 is false) and P(H_0 is true). The more plausible the research hypothesis is before we conduct the study, the less stringent the test need be before we reject H_0. However, if the research hypothesis has low prior probability of being true, then the study should be required to attain high statistical significance (i.e, a very low p-value) before any confidence is placed in it.

SUMMARY

1. The three steps in a significance test are: (1) determine a specific null hypothesis to be tested; (2) determine an appropriate test criterion—a statistic whose sampling distribution is known under the null hypothesis; and (3) calculate the test criterion from the sample data and determine the corresponding significance level.

2. A test of hypothesis differs from a significance test in that it entails predetermining a particular significance level to be used as a cutoff: if the p-value is larger than this significance level, the null hypothesis is accepted; otherwise it is rejected.

3. The p-value is the probability of observing what we actually did observe, or anything more extreme, if the null hypothesis is true. The result is significant at any level larger than or equal to p, but not significant at any level less than p. This is *statistical* significance, as opposed to *biological* or *medical* significance.

4. In a one-sided (one-tail) test, results that are more extreme in one direction only are included in the evaluation of p. In a two-sided test, results that are more extreme in both directions are included. Thus, to attain a specified significance level, the test statistic need be less "atypical" for a one-sided test than for a two-sided test.

5. Hypotheses about a proportion or percentile can be tested using the binomial distribution. For large samples, an observed proportion is about normally distributed with mean π and standard deviation $\sqrt{\pi(1 - \pi)/n}$.

6. To test the equality of two variances we use the F-statistic: the ratio of the two sample variances. The number of degrees of freedom in the top of the ratio corresponds to that along the top of the F-table, and the number in the bottom corresponds to that along the side of the table. For a two-sided test, the larger sample variance is put at the top of the ratio and the tail probability indicated by the table is doubled. The F-statistic is sensitive to nonnormality.

7. If we have normal distributions, the t-distribution can be used to test for the equality of two means. If we have two independent samples of sizes n_1 and n_2 from populations with the same variance, we use the two-sample t-test after estimating a pooled variance with $n_1 + n_2 - 2$ degrees of freedom. If we have a sample of n correlated pairs of observations, we use the n differences as a basis for the paired t-test, with $n - 1$ degrees of freedom.

8. If we have two populations with similarly shaped distributions, the rank sum test can be used to test the equality of the two medians when we have two independent samples, and the signed rank sum test when we have paired data. The sign test can also be used for paired data without making any assumption about the underlying distributions. All three tests are based on statistics that, when standardized, are about normally distributed in large samples.

9. A valid test is one for which the stated probability of the type 1 error (α) is correct: when the null hypothesis is true, it leads to rejection of the null hypothesis with probability α. A powerful test is one for which the probability of type II error (i.e., the probability of accepting the null hypothesis when it is false, β) is low.

10. The more assumptions that can be made, the more powerful a test can be. A one-sided test is more powerful than a two-sided test. The power of a statistical test can also be increased by using a larger significance level, a larger sample size, or by deciding to try to detect a larger difference; it is decreased by greater variability, whether due to measurement error or heterogeneity of study units.

FURTHER READING

Altman D.G. (1980) Statistics and Ethics in Medical Research: III. How Large a Sample? British Medical Journal 281:1336–1338. (This article contains a nomogram, for a two-sample t-test with equal numbers in each sample, relating power, total study size, the standardized mean difference, and the significance level. Given any three of these quantities, the fourth can be read off the nomogram.)

Blackwelder W.C. (1982) "Proving the Null Hypothesis" in Clinical Trials. Controlled Clinical Trials 3:345–353. (This article shows how to set up the statistical null hypothesis in a situation in which the research hypothesis of interest is that two different therapies are equivalent.)

Browner, W.S., and Newman, T.B. (1987) Are All Significant p-Values Created Equal? Journal of the American Medical Association 257:2459–2463. (This article develops in detail the analogy between diagnostic tests and tests of hypotheses.)

PROBLEMS

1. Significance testing and significance levels are important in the development of science because
   A. they allow one to prove a hypothesis is false
   B. they provide the most powerful method of testing hypotheses
   C. they allow one to quantify one's belief in a particular hypothesis other than the null hypothesis
   D. they allow one to quantify how unlikely a sample result is if the null hypothesis is false
   E. they allow one to quantify how unlikely a sample result is if the null hypothesis is true

2. A one-sided test to determine the significance level is particularly relevant for situations in which
   A. we have paired observations
   B. we know *a priori* the direction of any true difference
   C. only one sample is involved
   D. we have normally distributed random variables
   E. we are comparing just two samples

3. If a one-sided test indicates that the null hypothesis can be rejected at the 5% level, then
   A. the one-sided test is necessarily significant at the 1% level
   B. a two-sided test on the same set of data is necessarily significant at the 5% level

C. a two-sided test on the same set of data cannot be significant at the 5% level

D. a two-sided test on the same set of data is necessarily significant at the 10% level

E. the one-sided test cannot be significant at the 1% level

4. A researcher conducts a clinical trial to study the effectiveness of a new treatment in lowering blood pressure and concludes that "the lowering of mean blood pressure in the treatment group was significantly greater than that in the group on placebo ($p < 0.01$)." This means that

A. if the treatment has no effect, the probability of the treatment group having a lowering in mean blood pressure as great as or greater than that observed is exactly 1%

B. if the treatment has no effect, the probability of the treatment group having a lowering in mean blood pressure as great as or greater than that observed is less than 1%

C. there is exactly a 99% probability that the treatment lowers blood pressure

D. there is at least a 99% probability that the treatment lowers blood pressure

E. none of the above

5. A surgeon claims that at least three-quarters of his operations for gastric resection are successes. He consults a statistician and together they decide to conduct an experiment involving 10 patients. Assuming the binomial distribution is appropriate, the following probabilities are of interest:

Number of Successes	Probability of Success with $p = \frac{3}{4}$
0	0.0000
1	0.0000
2	0.0004
3	0.0031
4	0.0162
5	0.0582
6	0.1460
7	0.2503
8	0.2816
9	0.1877
10	0.0563

Suppose 4 of the 10 operations are successes. Which of the following conclusions is best?

A. The claim should be doubted, since the probability of observing 4 or fewer successes with $p = \frac{3}{4}$ is 0.0197.

B. The claim should not be doubted, since the probability of observing 4 or more successes is 0.965.
C. The claim should be doubted only if 10 successes are observed.
D. The claim should be doubted only if no successes are observed.
E. None of the above.

6. "The difference is significant at the 1% level" implies
A. there is a 99% probability that there is a real difference
B. there is at most a 99% probability of something as or more extreme than the observed result occurring if, in fact, the difference is zero
C. the difference is significant at the 5% level
D. the difference is significant at the 0.1% level
E. there is at most a 10% probability of a real difference

7. The p-value is
A. the probability of the null hypothesis being true
B. the probability of the null hypothesis being false
C. the probability of the test statistic or any more extreme result, assuming the null hypothesis is true
D. the probability of the test statistic or any more extreme result, assuming the null hypothesis is false
E. none of the above

8. A clinical trial is conducted to compare the efficacy of two treatments, A and B. The difference between the mean effects of the two treatments is not statistically significant. This failure to reject the null hypothesis could be because of all the following except
A. the sample size is large
B. the power of the statistical test is small
C. the difference between the therapies is small
D. the common variance is large
E. the probability of making a type II error is large

9. An investigator compared two weight-reducing agents and found the following results:

	Drug A	Drug B
Mean weight loss	10 lb	5 lb
Standard deviation	2 lb	1 lb
Sample size	16	16

Using a t-test, the p-value for testing the null hypothesis that the average reduction in weight was the same in the two groups was less than 0.001. An appropriate conclusion is

A. the sample sizes should have been larger
B. an F-test is called for
C. drug A appears to be more effective
D. drug B appears to be more effective
E. the difference between the drugs is not statistically significant

10. An investigator wishes to test the equality of the means of two random variables Y_1 and Y_2 based on a sample of matched pairs. It is known that the distribution of Y_1 is not normal but has the same shape as that of Y_2. Based on this information, the most appropriate test statistic in terms of validity and power is the

A. paired t-test
B. Wilcoxon's signed rank test
C. sign test
D. F-test
E. one-sided test

11. A lipid laboratory claimed it could determine serum cholesterol levels with a standard deviation no greater than that of a second laboratory. Samples of blood were taken from a series of patients. The blood was pooled, thoroughly mixed, and divided into aliquots. Twenty of these aliquots were labeled with fictitious names and ten sent to each laboratory for routine lipid analysis, interspersed with blood samples from other patients. Thus, the cholesterol determinations for these aliquots should have been identical, except for laboratory error. On examination of the data, the estimated standard deviations for the 10 aliquots were found to be 11 and 7 mg/dl for the first and second laboratories, respectively. Assuming cholesterol levels are approximately normally distributed, an F-test was performed of the null hypothesis that the standard deviation is the same in the two laboratories; it was found that $F = 1.57$ with 9 and 9 d.f. ($p \doteq 0.25$). An appropriate conclusion is

A. the data are consistent with the laboratory's claim
B. the data suggest the laboratory's claim is not valid
C. rather than an F-test, a t-test is needed to evaluate the claim
D. the data fail to shed light on the validity of the claim
E. a clinical trial would be more appropriate for evaluating the claim

12. We often make assumptions about data in order to justify the use of a specific statistical test procedure. If we say a test is robust to certain assumptions, we mean it

A. generates p-values having the desirable property of minimum variance
B. depends on the assumptions only through unbiased estimators
C. produces approximately valid results even if the assumptions are not true
D. is good only when the sample size exceeds 30
E. minimizes the chance of type II errors

13. A type II error is

A. the probability that the null hypothesis is true
B. the probability that the null hypothesis is false
C. made if the null hypothesis is accepted when it is false
D. made if the null hypothesis is rejected when it is true
E. none of the above

14. The power of a statistical test

A. should be investigated whenever a significant result is obtained
B. is a measure of significance
C. increases with the variance of the population
D. depends upon the sample size
E. should always be minimized

15. An investigator wishes to compare the ability of two competing statistical tests to declare a mean difference of 15 units statistically significant. The first test has probability 0.9 and the second test has probability 0.8 of being significant if the mean difference is in fact 15 units. It should be concluded for this purpose that

A. the first test is more powerful than the second
B. the first test is more robust than the second
C. the first test is more skewed than the second
D. the first test is more uniform than the second
E. the first test is more error prone than the second

CHAPTER EIGHT

The Many Uses of Chi-Square

CHAPTER EIGHT

Key Concepts

chi-square distribution
observed and expected values, goodness-of-fit tests
contingency table, dependent (response) variables, independent (factor) variables, association
McNemar's test for matched pairs
combining p-values
likelihood ratio criterion

Symbols and Abbreviations

$\log_e$	logarithm to base e; natural logarithm (''ln'' on many calculators)
x^2	sample chi-square statistic (also denoted X^2, χ^2)
χ^2	percentile of the chi-square distribution (also used to denote the corresponding statistic)

The Many Uses of Chi-Square

THE CHI-SQUARE DISTRIBUTION

In this chapter we introduce a distribution known as the **chi-square distribution,** denoted χ^2 (χ is the Greek letter chi, pronounced as "kite" without the "t" sound). This distribution has many important uses, including testing hypotheses about proportions and calculating confidence intervals for variances. Often you will read in the medical literature that "the chi-square test" was performed, or "chi-square analysis" was used, as though referring to a unique procedure. We shall see in this chapter that chi-square is not just a single method of analysis, but rather a distribution that is used in many different statistical procedures.

Suppose a random variable Y is known to have a normal distribution with mean μ and variance σ^2. We have stated that under these circumstances,

$$Z = \frac{Y - \mu}{\sigma}$$

has a standardized normal distribution (i.e., a normal distribution with mean zero and standard deviation one). Now the new random variable obtained by squaring Z, that is,

$$Z^2 = \frac{(Y - \mu)^2}{\sigma^2}$$

has a chi-square distribution with one degree of freedom. Only one random variable, Y is involved in Z^2, and there are no constraints; hence we say Z^2 has one degree of freedom. From now on we shall abbreviate "degrees of freedom" by d.f.: the distribution of Z^2 is chi-square with 1 d.f.

Whereas a normally distributed random variable can take on any value, positive or negative, a chi-square random variable can only be positive or zero (a squared number cannot be negative). Recall that about 68% of the distribution of Z (the standardized normal distribution) lies between −1 and +1; correspondingly, 68% of the distribution of Z^2, the chi-square distribution with 1 d.f., lies between 0 and +1. The remaining 32% lies between +1 and ∞. Therefore, the graph of χ^2 with 1 d.f. is positively skewed, as shown in Figure 8–1.

Suppose that Y_1 and Y_2 are independent random variables, each normally distributed with mean μ and variance σ^2. Then

$$Z_1^2 = \frac{(Y_1 - \mu)^2}{\sigma^2}$$

and

$$Z_2^2 = \frac{(Y_2 - \mu)^2}{\sigma^2}$$

are each distributed as χ^2 with one d.f. Moreover, $Z_1^2 + Z_2^2$ (i.e., the sum of these two independent, squared standardized normal random variables) is distributed as chi-square with 2 d.f. More generally, if we have a set of k independent random variables $Y_1, Y_2, \ldots, Y_k$, each normally distributed with mean μ and variance σ^2, then

$$\frac{(Y_1 - \mu)^2}{\sigma^2} + \frac{(Y_2 - \mu)^2}{\sigma^2} + \cdots + \frac{(Y_k - \mu)^2}{\sigma^2}$$

is distributed as χ^2 with k d.f.

Now consider replacing μ by its minimum variance unbiased estimator $\bar{Y}$, the sample mean. Once the sample mean $\bar{Y}$ is determined, there are $k - 1$ choices possible for the values of the Y's. Thus

$$\frac{(Y_1 - \bar{Y})^2}{\sigma^2} + \frac{(Y_2 - \bar{Y})^2}{\sigma^2} + \cdots + \frac{(Y_k - \bar{Y})^2}{\sigma^2}$$

is distributed as χ^2 with $k - 1$ d.f.

Figure 8–1 gives examples of what the chi-square distribution looks like. It is useful to remember that the mean of a chi-square distribution is equal to its number of d.f., and that its variance is equal to twice its number of d.f. Note also that as the number of d.f. increases, the distribution becomes less and less skewed; in fact, as a number of d.f. tends to infinity, the chi-square distribution tends to a normal distribution.

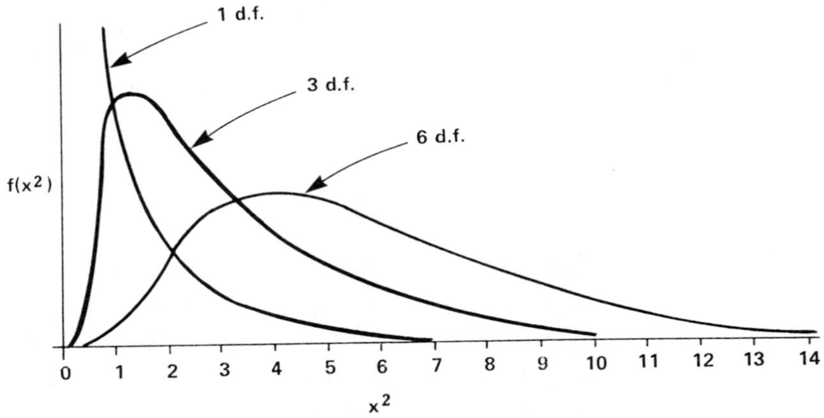

Figure 8–1. The chi-square distributions with 1, 3, and 6 d.f.

We now discuss some random variables that have approximate chi-square distributions. Let Y have an arbitrary distribution with mean μ and variance σ^2. Let $\overline{Y}$ be the mean of a random sample from the distribution. We know that for large samples

$$Z = \frac{\overline{Y} - \mu}{\sigma_{\overline{Y}}}$$

is approximately distributed as the standard normal, regardless of the shape of the distribution of Y. It follows that for large samples

$$Z^2 = \frac{(\overline{Y} - \mu)^2}{\sigma^2_{\overline{Y}}}$$

is approximately distributed as chi-square with 1 d.f., regardless of the shape of the distribution of Y. Similarly, if we sum k such quantities—each being the square of a standardized sample mean—then, provided they are independent, the sum is approximately distributed as chi-square, with k d.f.

Recall, for example, that if y is the outcome of a binomial random variable with parameters n and π, then in large samples the standardized variable

$$z = \frac{y/n - \pi}{\sqrt{\pi(1 - \pi)/n}} = \frac{y - n\pi}{\sqrt{n\pi(1 - \pi)}}$$

can be considered as the outcome of a standard normal random variable. Thus the square of this, z^2, can be considered to come from a chi-square distribution with 1 d.f. Now instead of writing y and $n - y$ for the numbers we observe in the two categories (e.g., the number affected and the number unaffected), let us write y_1 and y_2 so that $y_1 + y_2 = n$. Analogously let us write π_1 and π_2 so that $\pi_1 + \pi_2 = 1$. Then

$$z^2 = \frac{(y_1 - n\pi_1)^2}{n\pi_1(1 - \pi_1)} = \frac{(y_1 - n\pi_1)^2}{n\pi_1} + \frac{(y_2 - n\pi_2)^2}{n\pi_2}$$

(If you wish to follow the steps that show this, see Appendix 1). Now notice that each of the two terms on the right corresponds to one of the two possible outcomes for a binomially distributed random variable: y_1 and $n\pi_1$ are the observed and expected numbers in the first category (affected), and y_2 and $n\pi_2$ are the observed and expected numbers in the second category (unaffected). Each term is of the form

$$\frac{(\text{observed} - \text{expected})^2}{\text{expected}}$$

Adding them together, we obtain a statistic that comes from (approximately in large samples) a chi-square distribution with 1 d.f. We shall see how this result can be generalized to generate many different "chi-square statistics," often called "Pearson chi-squares" after the British statistician Karl Pearson (1857–1936).

GOODNESS-OF-FIT TESTS

To illustrate the use of this statistic, consider the offspring of matings in which one parent is hypercholesterolemic and the other is normocholesterolemic. We wish to test the hypothesis that children from such matings are observed in the 1:1 ratio of hypercholesterolemic to normocholesterolemic, as expected from matings of this type if hypercholesterolemia is a rare autosomal dominant trait. Thus we observe a set of children from such matings and wish to test the "goodness of fit" of the data to the hypothesis of autosomal dominant inheritance. A 1:1 ratio implies probabilities $\pi_1 = \pi_2 = 0.5$ for each category, and this will be our null hypothesis, H_0. Suppose the observed numbers are 87 and 79, so that $n = 87 + 79 = 166$. The expected numbers under H_0 are $n\pi_1 = 166(0.5) = 83$, and $n\pi_2 = 166(0.5) = 83$. The chi-square statistic is thus

$$x^2 = \frac{(87 - 83)^2}{83} + \frac{(79 - 83)^2}{83} = 0.39$$

(From now on we shall use x^2 to denote the observed value of a chi-square statistic.) If we had observed the number of children expected under H_0 in each category (i.e., 83), the chi-square statistic x^2, would be zero. Departure from this in either direction (either too many hypercholesterolemic or too many normocholesterolemic offspring) increases x^2. Thus we reject H_0 if x^2 is large (i.e., above the 95th percentile of the chi-square distribution with a 1 d.f. for a test at the 5% significance level). The p-value is the area of the chi-square distribution above the observed x^2, and this automatically allows for departure from H_0 in either direction. From Table 4 in Appendix 2, we see that the 95th percentile of the chi-square distribution with 1 d.f. is 3.84. Since 0.39 is less than 3.84, the departure from H_0 is not significant at the 5% level. In fact 0.39 corresponds to $p = 0.54$, so that the fit to autosomal dominant inheritance is very good.

This test can easily be generalized to any number of categories. Suppose we have a sample of *n* observations, and each observation must fall in one, and only one, of k possible categories. Denote the numbers that are **observed** in each category $o_1, o_2, \ldots, o_k$, and the corresponding numbers that are **expected** (under a particular H_0) $e_1, e_2, \ldots, e_k$. Then the chi-square statistic is simply

$$x^2 = \frac{(o_1 - e_1)^2}{e_1} + \frac{(o_2 - e_2)^2}{e_2} + \ldots + \frac{(o_k - e_k)^2}{e_k}$$

and under H_0, this can be considered as coming from a chi-square distribution with $k - 1$ d.f. (Once the total number of observations, *n*, is fixed, arbitrary numbers can be placed in only $k - 1$ categories). Of course, the sample size must be large enough. The same rule of thumb that we have introduced before can be used to check this: if each expected value is at least 5, the chi-square approximation is good. The approximation may still be good if a few of the expected values are less than 5, but in that situation it is common practice to pool two or more of the categories with small expected values.

As an example with three categories, consider the offspring of parents whose red cells agglutinate when mixed with either anti-M or anti-N sera. If these reactions are detecting two alleles at a single locus, then the parents are heterozygous (MN). Furthermore, the children should be MM (i.e., their red cells agglutinate only with anti-M) with probability 0.25, MN (like their parents) with probability 0.5, or NN (i.e., their cells agglutinate only with anti-N) with probability 0.25. Suppose we test the bloods of 200 children and observe in the three categories: $o_1 = 42$, $o_2 = 106$, and $o_3 = 52$, respectively. To test how well these data fit the hypothesis of two alleles segregating at a single locus, we calculate the appropriate chi-square statistic, with $e_1 = 200(0.25) = 50$, $e_2 = 200(0.5) = 100$, and $e_3 = 200(0.25) = 50$. The computations can be conveniently arranged as in Table 8–1. In this case we compare 1.72 to the chi-square distribution with 2 ($k - 1 = 3 - 1 = 2$) d.f. Table 4 in Appendix 2 shows that the 95th percentile of the distribution is 5.99, and so the departure from what is expected is not significant at the 5% level. In fact $p = 0.43$, and once again the fit is good.

The same kind of test can be used to determine the **goodness-of-fit** of a set of sample data to any distribution. We mentioned in the last chapter that there are tests to determine if a set of data could reasonably come from a normal distribution, and this is one such test. Suppose, for example, we wanted to test the goodness-of-fit of the serum cholesterol levels in Table 3.2 to a normal distribution. The table gives the observed numbers in 20 different categories. We obtain the expected numbers in each category on the basis of the best-fitting normal distribution, substituting the sample mean and variance for the population values. In this way we have 20 observed numbers and 20 expected numbers, and so can obtain x^2 as the sum of 20 components. Because we force the expected number to come from a distribution with exactly the same mean and variance as in the sample, however, there are two fewer degrees of freedom. Thus in this case we would compare x^2 to the chi-square distribution with $20 - 1 - 2 = 17$ d.f. Note, however, that in the extreme categories of Table 3–2 there are some small numbers. If any of these categories have expected numbers below five, it might be necessary to pool them with neighboring categories. The total number of categories, and hence also the number of d.f., would then be further reduced.

TABLE 8–1. **Computation of the Chi-Square Statistic to Test Whether the MN Phenotypes of 200 Offspring of MN × MN Matings Are Consistent with a Two-Allele, One-Locus Hypothesis**

Hypothesized Genotype	Number Observed (o)	Number Expected (e)	(o − e)	Contribution to x^2 $[(o - e)^2/e]$
MM	42	50	−8	1.28
MN	106	100	+6	0.36
NN	52	50	+2	0.08
Total	200	200	0	$x^2 = 1.72$

CONTINGENCY TABLES

Categorical data are often arranged in a table of rows and columns in which the individual entries are counts. Such a table is called a **contingency table.** Two qualitative variables are involved, the rows representing the categories of the first variable and the columns representing the categories of the second variable. Each cell in the table represents a combination of categories of the two variables. Each entry in a cell is the number of study units observed in that combination of categories. Tables 8–2 and 8–3 are each examples of contingency tables with two rows and two columns (the row and column indicating totals are not counted). We call these two-way tables. In any two-way table, the hypothesis of interest—and hence the choice of an appropriate test statistic—is related to the types of variables being studied. We distinguish between two types of variables: dependent and independent variables. A **dependent,** or **response, variable** is one for which the distribution of study units in the different categories is merely observed by the investigator. A dependent variable is also sometimes

TABLE 8–2. **215 Medical Students Classified by Class and Serum Cholesterol Level (Above or Below 210 mg/dl)**

Class	Cholesterol Level: Normal	High	Total
First year	75	35	110
Fourth year	95	10	105
Total	170	45	215

TABLE 8–3. **110 First-year Medical Students Classified by Serum Cholesterol Level (Above or Below 210 mg/dl) and Serum Triglyceride Level (Above or Below 150 mg/dl)**

Cholesterol Level	Triglyceride Level: Normal	High	Total
Normal	60	15	75
High	20	15	35
Total	80	30	110

called a criterion variable or variate. An **independent,** or **factor, variable** is one for which the investigator actively controls the distribution of study units in the different categories. An independent variable is also sometimes called a predictor variable. Notice that we are using the term "independent" with a different meaning from that used in our discussion of probability. Although it is less confusing to use the terms "response" and "factor" variables, the terminology "dependent" and "independent" to describe these two different types of variables is so widely used that we shall use it too.

We shall consider the following two kinds of contingency tables:

1. The two-sample, one-independent-variable and one-dependent-variable case; the rows will be categories of an independent variable, whereas the columns will be categories of a dependent variable.

2. The one-sample, two-dependent-variables case; the rows will be categories of one dependent variable and the columns will be categories of a second dependent variable.

CASE 1

Consider the data in Table 8–2. Ignoring the row and column of totals, this table has four cells. It is known as a fourfold table, or a 2×2 contingency table. The investigator took one sample of 110 first-year students and a second sample of 105 fourth-year students; these are two categories of an independent variable, because the investigator had control over how many first- and fourth-year students were taken. A blood sample was drawn from each student and analyzed for cholesterol level; each student was then classified as having normal or high cholesterol level based on a prespecified cut-point. These are two categories of a dependent variable, because the investigator observed, rather than controlled, how many students fell in each category. Thus, in this example, student class is an independent variable and cholesterol level is a dependent variable.

Bearing in mind the way the data were obtained, we can view them as having the underlying probability structure shown in Table 8–4. Note that the first subscript of each π indicates the row it is in, and the second indicates the column it is in. Note also that $\pi_{11} + \pi_{12} = 1$ and $\pi_{21} + \pi_{22} = 1$.

Suppose we are interested in testing the null hypothesis H_0 that the proportion of the first year class with high cholesterol levels is the same as that of the fourth year class (i.e., $\pi_{12} = \pi_{22}$). (If this is true, it follows automatically that $\pi_{11} = \pi_{21}$ as well.) Assuming H_0 is true, we can estimate the "expected" number in each cell from the overall proportion of students who have normal or high cholesterol levels (i.e., from the proportions 170/215 and 45/215, respectively; see the last line of Table 8–2). In other words, if $\pi_{11} = \pi_{21}$ (which we shall then call

TABLE 8–4. **Probability Structure for the Data in Table 8–2**

Class	Cholesterol Level: Normal	High	Total
First year	π_{11}	π_{12}	1
Fourth year	π_{21}	π_{22}	1
Total	$\pi_{11} + \pi_{21}$	$\pi_{12} + \pi_{22}$	2

π_1), and $\pi_{12} = \pi_{22}$ (which we shall then call π_2), we can think of the 215 students as forming a single sample and the probability structure indicated in the "Total" row of Table 8–4 becomes

	π_1	π_2	1
instead of	$\pi_{11} + \pi_{21}$	$\pi_{12} + \pi_{22}$	2

We estimate π_1 by $p_1 = 170/215$ and π_2 by $p_2 = 45/215$. Now denote the total number of first-year students n_1 and the total number of fourth year students n_2. Then the expected numbers in each cell of the table are (always using the same convention for the subscripts—first one indicates the row, second one indicates the column):

$$\begin{aligned} e_{11} &= n_1p_1 = 110(170/215) = 86.98 \\ e_{12} &= n_1p_2 = 110(45/215) = 23.02 \\ e_{21} &= n_2p_1 = 105(170/215) = 83.02 \\ e_{22} &= n_2p_2 = 105(45/215) = 21.98. \end{aligned}$$

Note that the expected number in each cell is the product of the marginal totals corresponding to that cell, divided by the grand total. The chi-square statistic is then given by

$$\begin{aligned} x^2 &= \frac{(o_{11} - e_{11})^2}{e_{11}} + \frac{(o_{12} - e_{12})^2}{e_{12}} + \frac{(o_{21} - e_{21})^2}{e_{21}} + \frac{(o_{22} - e_{22})^2}{e_{22}} \\ &= \frac{(75 - 86.98)^2}{86.98} + \frac{(35 - 23.02)^2}{23.02} + \frac{(95 - 83.02)^2}{83.02} + \frac{(10 - 21.98)^2}{21.98} \\ &= 16.14. \end{aligned}$$

The computations can be conveniently arranged in tabular form, as indicated in Table 8–5.

In this case the chi-square statistic equals (about) the square of a single standardized normal random variable, and so has 1 d.f. Intuitively, we can deduce the number of d.f. by noting that we used the marginal totals to estimate the expected numbers in each cell, so that we forced the marginal totals of the expected numbers to equal the marginal totals of the observed numbers. (Check that this is so.) Now if we fix all the marginal totals, how many of the cells of the

TABLE 8–5. **Computation of the Chi-Square Statistic for the Data in Table 8–2**

Class	Cholesterol Level	Number Observed (o)	Number Expected (e)	o − e	Contribution to x^2 $[(o - e)^2/e]$
First year	Normal	75	86.98	−11.98	1.65
	High	35	23.02	11.98	6.23
Fourth year	Normal	95	83.02	11.98	1.73
	High	10	21.98	−11.98	6.53
Total		215	215.00		$x^2 = 16.14$

2 × 2 table can be filled in with arbitrary numbers? The answer is only one; once we fill a single cell of the 2 × 2 table with an arbitrary number, that number and the marginal totals completely determine the other three entries in the table. Thus there is only 1 d.f. Looking up the percentile values of the chi-square distribution with 1 d.f. (Table 4, Appendix 2), we find that the 99.9th percentile is 10.83. Since 16.14 is greater than this, the two proportions are significantly different at the 0.1% level (i.e., $p < 0.001$). We conclude that the proportion with high cholesterol levels is significantly different for first-year and fourth-year students. Equivalently, we conclude that the distribution of cholesterol levels depends on (is associated with) the class to which a student belongs, or that the two variables student class and cholesterol level are not independent in the probability sense. (As indicated earlier, this statement has nothing to do with our calling student class an independent variable and cholesterol level a dependent variable.)

CASE 2

Consider now the data given in Table 8–3. Here we have a single sample of 110 first-year medical students and have observed whether each student is high or normal, with respect to specified cut-points, for two dependent variables: cholesterol level and triglyceride level. These data can be viewed as having the underlying probability structure shown in Table 8–6, which should be con-

TABLE 8–6. **Probability Structure for the Data in Table 8–3**

	Triglyceride Level		
Cholesterol Level	Normal	High	Total
Normal	π_{11}	π_{12}	$\pi_{1.}$
High	π_{21}	π_{22}	$\pi_{2.}$
Total	$\pi_{.1}$	$\pi_{.2}$	1

trasted with Table 8–4. Notice that dots are used in the marginal totals of Table 8–6 (e.g., $\pi_{1.} = \pi_{11} + \pi_{12}$); so that a dot replacing a subscript indicates that the π is the sum of the π's with different values of that subscript.

Suppose we are interested in testing the null hypothesis H_0 that triglyceride level is not associated with cholesterol level (i.e., triglyceride level is independent of cholesterol level in a probability sense). Recalling the definition of independence from Chapter 4, we can state H_0 as being equivalent to

$$\pi_{11} = \pi_{1.}\pi_{.1}$$
$$\pi_{12} = \pi_{1.}\pi_{.2}$$
$$\pi_{21} = \pi_{2.}\pi_{.1}$$
$$\pi_{22} = \pi_{2.}\pi_{.2}$$

Assuming H_0 is true, we can once again estimate the expected number in each cell of the table. We first estimate the marginal proportions of the table. Using the letter p to denote an estimate of the probability, these are

$$p_{1.} = \frac{75}{110}$$
$$p_{2.} = \frac{35}{110}$$
$$p_{.1} = \frac{80}{110}$$
$$p_{.2} = \frac{30}{110}$$

Then each cell probability is estimated as a product of the two corresponding marginal probabilities (because if H_0 is true we have independence). Thus, letting n denote the total sample size, under H_0 the expected numbers are calculated to be

$$e_{11} = np_{1.}p_{.1} = 110\left(\frac{75}{110}\right)\left(\frac{80}{110}\right) = 54.55$$
$$e_{12} = np_{1.}p_{.2} = 110\left(\frac{75}{110}\right)\left(\frac{30}{110}\right) = 20.45$$
$$e_{21} = np_{2.}p_{.1} = 110\left(\frac{35}{110}\right)\left(\frac{80}{110}\right) = 25.45$$
$$e_{22} = np_{2.}p_{.2} = 110\left(\frac{35}{110}\right)\left(\frac{30}{110}\right) = 9.55$$

Note that after canceling out 110, each expected number is once again the product of the two corresponding marginal totals divided by the grand total. Thus we can calculate the chi-square statistic in exactly the same manner as before, and once again the resulting chi-square has 1 d.f. Table 8–7 summarizes the calculations. The calculated value, 6.27, lies between the 98th and 99th percentiles (5.41 and 6.64, respectively: see Table 4, Appendix 2) of the chi-square distri-

TABLE 8–7. **Computation of the Chi-Square Statistic for the Data in Table 8–3**

Cholesterol Level	Triglyceride Level	Number Observed (o)	Number Expected (e)	o − e	Contribution to x^2 $[(o - e)^2/e]$
Normal	Normal	60	54.55	+5.45	0.54
	High	15	20.45	−5.45	1.45
High	Normal	20	25.45	−5.45	1.17
	High	15	9.55	+5.45	3.11
Total		110	110.00		$x^2 = 6.27$

bution with 1 d.f. We therefore conclude that triglyceride levels and cholesterol levels are not independent but are associated ($0.01 < p < 0.02$).

Suppose now that we ask a different question, again to be answered using the data in Table 8–3: Is the proportion of students with high cholesterol levels different from the proportion with high triglyceride levels? In other words, we ask whether the two dependent variables, dichotomized, follow the same binomial distribution. Our null hypothesis H_0 is that the two proportions are the same; i.e.,

$$\pi_{21} + \pi_{22} = \pi_{12} + \pi_{22}$$

which is equivalent to $\pi_{21} = \pi_{12}$.

A total of $20 + 15 = 35$ students are in the two corresponding cells, and under H_0 the expected number in each would be half this, that is,

$$e_{12} = e_{21} = \frac{1}{2}(o_{12} + o_{21}) = \frac{35}{2} = 17.5$$

The appropriate chi-square statistic to test this hypothesis is thus

$$x^2 = \frac{(o_{12} - e_{12})^2}{e_{12}} + \frac{(o_{21} - e_{21})^2}{e_{21}} = \frac{(20 - 17.5)^2}{17.5} + \frac{(15 - 17.5)^2}{17.5} = 0.71$$

The numbers in the other two cells of the table are not relevant to the question asked, and so the chi-square statistic for this situation is formally the same as the one we calculated earlier to test for mendelian proportions among offspring of one hypercholesterolemic and one normocholesterolemic parent. Once again it has 1 d.f. and there is no significant difference at the 5% level (in fact, $p = 0.4$).

Regardless of whether or not it would make any sense, we cannot apply the probability structure in Table 8–6 to the data in Table 8–2 and analogously ask whether $\pi_{.1}$ and $\pi_{1.}$ are equal (i.e., is the proportion of fourth-year students equal to the proportion of students with high cholesterol?). The proportion of students who are fourth-year cannot be estimated from the data in Table 8–2, because we were told that the investigator took a sample of 110 first-year students and a sample of 105 fourth-year students. The investigator had complete control over how many students in each class came to be sampled, regardless of how many there happened to be. If, on the other hand, we had been told that a random sample of all first- and fourth-year students had been taken, and it was then observed that the sample contained 110 first-year and 105 fourth-year students, then student class would be a dependent variable and we could test the null hypothesis $\pi_{.1} = \pi_{1.}$.

You might think that any hypothesis of this nature is somewhat artificial; after all, whether or not the proportion of students with high cholesterol is equal to the proportion with high triglyceride is merely a reflection of the cut-points used for each variable. There is, however, a special situation where this kind of question, requiring the test we have just described (which is called **McNemar's test**), often occurs. Suppose we wish to know whether the proportion of men and women with high cholesterol levels is the same, for which we would naturally define "high" by the same cut-point in the two sexes. One way to do this would be to take two samples—one of men and one of women—and perform the first test we described for a 2×2 contingency table (Case 1). The situation would be analogous to that summarized in Table 8–2, the independent variable being gender rather than class. But cholesterol levels change with age. Unless we take the precaution of having the same age distribution in each sample, any difference that is found could be due to either the age difference or the sex difference between the two groups. For this reason it would be wise to take a sample of matched pairs, each pair consisting of a man and a woman of the same age. If we have n such pairs, we do not have $2n$ independent individuals, because of the matching. By considering each *pair* as a study unit, however, it is reasonable to suppose that we have n independent study units, with two different dependent variables measured on each—cholesterol level of the woman of the pair and cholesterol level of the man of the pair. We would then draw up a table similar to Table 8–3, but with each *pair* as a study unit. Thus, corresponding to the 110 medical students, we would have n, the number of pairs; and the two dependent variables, instead of cholesterol and triglyceride level, would be cholesterol level in the woman of each pair and cholesterol level in the man of each pair. To test whether the proportion with high cholesterol is the same in men and women, we would now use McNemar's test, which uses the information in only those two cells of the 2×2 table that relate to untied pairs.

There are thus two different ways in which we could conduct a study to answer the same question: Is cholesterol level independent of sex?

Because of the different ways in which the data are sampled, two different chi-square tests are necessary: the first is the usual contingency-table chi-square test, the second is McNemar's test.

In general, a two-way contingency table can have any number r of rows and any number c of columns, and the contingency table chi-square is used to test whether the row variable is independent of or associated with the column variable. The general procedure is to use the marginal totals to calculate an "expected" number for each cell, and then to sum the quantities (observed − expected)^2/expected for all r × c cells. Fixing the marginal totals, it is found that (r − 1)(c − 1) cells can be filled in with arbitrary numbers, and so this chi-square has (r − 1)(c − 1) degrees of freedom. When r = c = 2 (the 2 × 2 table), (r − 1)(c − 1) = (2 − 1)(2 − 1) = 1. For the resulting statistic to be distributed as chi-square under H_0, we must

1. have each study unit appearing in only one cell of the table
2. sum the contributions over all the cells, so that all the study units in the sample are included
3. have in each cell the count of a number of *independent* events
4. not have small expected values causing large contributions to the chi-square.

Note conditions 3 and 4. Suppose our study units are children and these have been classified according to disease status. If disease status is in any way familial, then two children in the same family are not independent. Although condition 3 would be satisfied if the table contains only one child per family, it would not be satisfied if sets of brothers and sisters are included in the table. In such a situation the "chi-square" statistic would not be expected to follow a chi-square distribution. With regard to condition 4, it is sufficient for the expected value in each cell to be at least 5. If this condition does not hold, the chi-square statistic may be spuriously large and for such a small sample it may be necessary to use a test known as Fisher's exact test.

Before leaving the subject of contingency tables, a cautionary note is in order regarding the interpretation of any significant dependence or **association** that is found. As stated in Chapter 4, many different causes may underlie the dependence between two events. Consider, for example, the following fourfold table, in which 2,000 persons are classified as possessing a particular antigen (A+) or not (A−), and as having a particular disease (D+) or not (D−):

	A+	A−	Total
D+	51	59	110
D−	549	1,341	1,890
Total	600	1,400	2,000

We see that among those persons who have the antigen, 51/600 = 8.5% have the disease, whereas among those who do not have the antigen, 59/1,400 = 4.2% have the disease. There is a clear association between the two variables, which is highly significant (chi-square with 1 d.f. = 14.84, $p < 0.001$). Does this mean that possession of the antigen predisposes to having the disease? Or that having the disease predisposes to possession of the antigen? Neither of these interpretations may be correct, as we shall see.

Consider the following two analogous fourfold tables, one pertaining to 1,000 white persons and one pertaining to 1,000 black persons:

	Whites				Blacks		
	A+	A−	Total		A+	A−	Total
D+	50	50	100	D+	1	9	10
D−	450	450	900	D−	99	891	990
Total	500	500	1,000	Total	100	900	1,000

In neither table is there any association between possession of the antigen and having the disease. The disease occurs among 10% of the white persons, whether or not they possess the antigen; and it occurs among 1% of the black persons, whether or not they possess the antigen. The antigen is also more prevalent in the white sample than in the black sample. Because of this, when we add the two samples together—which results in the original table for all 2,000 persons—a significant association is found between possession of the antigen and having the disease. Thus an association can be caused merely by mixing samples from two or more populations, or by sampling from a single heterogeneous population. Populations may be heterogeneous with respect to race, age, sex, or any number of other factors that could be the cause of an association. A sample should always be stratified with respect to such factors before performing a chi-square test for association. Then either the test for association should be performed separately within each stratum, or an overall statistic used that specifically tests for association *within* the strata. One such overall test statistic often referred to in the medical literature is the Mantel-Haenszel chi-square.

INFERENCE ABOUT THE VARIANCE

Let us suppose we have a sample $Y_1, Y_2, \ldots, Y_n$ from a normal distribution with variance σ^2, and let the sample mean be $\bar{Y}$. We have seen that

$$\frac{(Y_1 - \bar{Y})^2}{\sigma^2} + \frac{(Y_2 - \bar{Y})^2}{\sigma^2} + \ldots + \frac{(Y_n - \bar{Y})^2}{\sigma^2}$$

then follows a chi-square distribution with $n - 1$ degrees of freedom. But this expression can also be written in the form

$$\frac{(n - 1)S^2}{\sigma^2}$$

where S^2 is the sample variance. Thus, denoting the 2.5th and 97.5th percentiles of the chi-square distribution with $n - 1$ d.f. as $\chi^2_{2.5}$ and $\chi^2_{97.5}$, respectively, we know that

$$P\left(\chi^2_{2.5} \leq \frac{(n - 1)S^2}{\sigma^2} \leq \chi^2_{97.5}\right) = 0.95$$

This statement can be written in the equivalent form

$$P\left(\frac{(n - 1)S^2}{\chi^2_{97.5}} \leq \sigma^2 \leq \frac{(n - 1)S^2}{\chi^2_{2.5}}\right) = 0.95$$

which gives us a way of obtaining a 95% confidence interval for a variance; all we need do is substitute the specific numerical value s^2 from our sample in place of the random variable S^2. In other words, we have 95% confidence that the true variance σ^2 lies between the two numbers

$$\frac{(n - 1)s^2}{\chi^2_{97.5}} \quad \text{and} \quad \frac{(n - 1)s^2}{\chi^2_{2.5}}$$

For example, suppose we found $s^2 = 4$ with 10 d.f. Table 4 in Appendix 2 does not give the 2.5th and 97.5th percentiles, so we shall use the 5th and

95th percentiles of the chi-square distribution with 10 d.f. (3.94 and 18.31) to obtain a 90% confidence interval. The limits of the interval are

$$\frac{(10)4}{18.31} = 2.18 \qquad \text{and} \qquad \frac{(10)4}{3.94} = 10.15$$

Notice that even though this is a 90% confidence interval (which is necessarily narrower than a 95% confidence interval), it is nevertheless quite wide. Typically we require a much larger sample to estimate a variance or standard deviation than we require to estimate, with the same degree of precision, a mean.

We can also test hypotheses about a variance. If we wanted to test the hypothesis $\sigma^2 = 6$ in the above example, we would compute

$$x^2 = \frac{(10)(4)}{6} = 6.67$$

which, if the hypothesis is true, comes from a chi-square distribution with 10 degrees of freedom. Since 6.67 is between the 70th and 80th percentiles of that distribution, there is no evidence to reject the hypothesis.

We have already discussed the circumstances under which the F-test can be used to test the hypothesis that two population variances are equal. Although the details are beyond the scope of this book, you should be aware of the fact that it is possible to test for the equality of a set of more than two variances, and that at least one of the tests to do this is based on the chi-square distribution. Remember, however, that all chi-square procedures for making inferences about variances depend rather strongly on the assumption of normality; they are quite sensitive to nonnormality of the underlying variable.

COMBINING p-VALUES

Suppose five investigators have conducted different experiments to test the same null hypothesis H_0 (for example, that two treatments have the same effect). Suppose further that the tests of significance of H_0 (that the mean response to treatment is the same) resulted in the p-values: $p_1 = 0.15$, $p_2 = 0.07$, $p_3 = 0.50$, $p_4 = 0.22$, and $p_5 = 0.09$. At first glance you might conclude that there is no significant difference between the two treatments. There is a way of pooling p-values from separate investigations, however, to obtain an overall p-value.

For any arbitrary p-value, if the null hypothesis that gives rise to it is true, $-2 \log_e p$ can be considered as coming from the χ^2 distribution with 2 d.f. ($\log_e$ stands for "logarithm to base e," or natural logarithm; it is denoted ln on many calculators). If there are k independent investigations, the corresponding p-values will be independent. Thus the sum of k such values,

$$-2 \log_e p_1 - 2 \log_e p_2 \ldots -2 \log_e p_k$$

can be considered as coming from the χ^2 distribution with 2k d.f. Hence, in the above example, we would calculate

$$\begin{aligned} -2 \log_e (0.15) &= 3.79 \\ -2 \log_e (0.07) &= 5.32 \\ -2 \log_e (0.50) &= 1.39 \\ -2 \log_e (0.22) &= 3.03 \\ -2 \log_e (0.09) &= \underline{4.82} \\ \text{Total} &= 18.35 \end{aligned}$$

If H_0 is true, 18.35 comes from a χ^2 distribution with $2k = 2(5) = 10$ d.f. From Table 4, Appendix 2, we see that for the distribution with 10 d.f., 18.31 corresponds to $p = 0.05$. Thus, by pooling the results of all five investigations, we see that the treatment difference is just significant at the 5% level. It is, of course, necessary to check that each investigator is in fact testing the same null hypothesis.

LIKELIHOOD RATIO TESTS

Earlier we introduced maximum likelihood estimators as a type of estimator with many good properties. Since different statistical tests compete with each other in much the same way that different estimators do, it would be helpful to have a general approach to derive statistical tests that have desirable properties. One such approach is to use a test based on the **likelihood ratio criterion.**

Recall that we ended Chapter 4 by calling the ratio of two conditional probabilities a likelihood ratio. In that instance, we were discussing the probability of a child receiving a B allele conditional on the child's father being a man accused of being the father, relative to the probability of the same event conditional on the child's father being a random man from a particular population. We can think of these two possibilities as two different hypotheses. In general, the likelihood ratio criterion is simply the likelihood for a set of data under one hypothesis divided by the likelihood for the same set of data under a second hypothesis. In either case, however, the likelihood may depend upon unknown parameters. When this occurs we simply substitute the maximum likelihood estimates for the parameters. In other words, each likelihood is maximized (mathematically or numerically) over any unknown parameters. Suppose, for example, we have two random samples. Under one hypothesis (the null hypothesis), they both come from identical normal distributions—same means and variances; under the other, alternative, hypothesis they come from two normal distributions with the same variances but different means. We wish to test the null hypothesis that the means of the two samples are not different. (Of course, we already know that the two-sample t-test is appropriate for this situation, but we shall nevertheless use it to illustrate how the likelihood ratio test works). Under each hypothesis the likelihood depends on the mean(s) and the variance, which are unknown parameters. So, to obtain the likelihood ratio, we maximize the likelihood twice: once over all possible values for a common mean and a common variance (i.e., under the null hypothesis), and once over all possible values for separate sample means and a common variance (the alternative hypothesis). The likelihood ratio criterion is then the maximum likelihood under the null hypothesis divided by the maximum likelihood under the alternative hypothesis.

The beauty of the likelihood ratio criterion lies in the following general theorem that holds under certain well-defined conditions: As the sample size increases, $-2 \log_e$(likelihood ratio), i.e., minus twice the natural logarithm of the likelihood ratio, tends to be distributed as chi-square if the null hypothesis is true. Thus, given large samples, if the two population means in our example are identical, $-2 \log_e$(likelihood ratio) is approximately distributed as chi-square. The number of d.f. is equal to the number of restrictions implied by the null hypothesis. In our example, the null hypothesis is that the two means are equal, which is a single restriction, so there is 1 d.f. For this particular example the test becomes identical to the two-sample t-test as the sample size tends to infinity. In fact, most of the tests discussed in this book are "asymptotically" identical to a test based on the likelihood ratio criterion. In those cases in which it has not been mathematically possible to derive an exact test, this general test based on a chi-square distribution is often used. Since it is now feasible, with modern computers, to calculate likelihoods corresponding to very elaborate probability models, this general approach is becoming more common in the medical literature. We shall discuss some examples in Chapter 11.

SUMMARY

1. Chi-square is a family of distributions used in many statistical procedures. Theoretically, the chi-square distribution with k d.f. is the distribution of the sum of k independent random variables, each of which is the square of a standardized normal random variable.

2. In practice we often sum more than k quantities that are not independent, but the sum is in fact equivalent to the sum of k independent quantities. The integer k is then the number of degrees of freedom (d.f.) associated with the chi-square distribution. In most situations there is an intuitive way of determining the number of d.f. When the data are counts, we often sum quantities of the form (observed − expected)^2/expected; the number of d.f. is then the number of counts that could have been arbitrarily chosen—with the stipulation that there is no change in the total number of counts or other specified parameters. Large values of the chi-square statistic indicate departure from the null hypothesis.

3. A chi-square goodness-of-fit test can be used to test whether a sample of data is consistent with any specified probability distribution. In the case of continuous traits, the data are first categorized in the manner used to construct a histogram. Categories with small expected numbers (less than 5) are usually pooled into larger categories.

4. In a two-way contingency table, either or both of the row and column variables may be dependent (response) variables. One variable may be controlled by the investigator and is then called an independent (factor) variable.

5. The hypothesis of interest determines which chi-square test is performed. Association, or lack of independence between two variables, is tested by the usual contingency table chi-square. The expected number in each cell is obtained as the product of the corresponding row and column totals divided by the grand total. The number of

degrees of freedom is equal to the product: (number of rows − 1)(number of columns − 1). Each study unit must appear only once in the table, and each count within a cell must be the count of a number of independent events.

6. For a 2 × 2 table in which both rows and columns are dependent variables, McNemar's test is used to test whether the two variables follow the same binomial distribution. If the study units are matched pairs (e.g., men and women matched by age), and each variable is a particular measurement on a specific member of the pair (e.g., cholesterol level on the man of the pair and cholesterol level on the woman of the pair), then McNemar's test is used to test whether the binomial distribution (low or high cholesterol level) is the same for the two members of the pair (men and women). This tests whether the specific measurement (cholesterol level) is independent of the member of the pair (gender).

7. The chi-square distribution can be used to construct a confidence interval for the variance of a normal random variable, or to test that a variance is equal to a specified quantity. This interval and this test are not robust against non-normality.

8. A set of p-values that result from independent investigations, all testing the same null hypothesis, can be combined to give an overall test of the null hypothesis. The sum of k independent quantities, $-2 \log_e p$, is compared to the chi-square distribution with 2k d.f.; a significantly large chi-square suggests that overall, the null hypothesis is not true.

9. The likelihood ratio criterion provides a general method of testing a hypothesis in large samples. Many of the usual statistical tests become identical to a test based on the likelihood ratio criterion as the sample size becomes infinite. The likelihood ratio criterion is the maximum likelihood for a particular sample of data under one (the null) hypothesis divided by the maximum likelihood for the same sample of data under another (alternative) hypothesis. Under certain well defined conditions, $-2 \log_e$ (likelihood ratio) can be used to test the null hypothesis, for it is approximately distributed as chi-square. The number of d.f. is the same as the number of restrictions implied by the null hypothesis.

FURTHER READING

Everitt, B.S. (1991) Analysis of Contingency Tables, 2nd ed. Chapman and Hall, London and New York.

PROBLEMS

1. The chi-square distribution is useful in all the following except
   A. testing the equality of two proportions
   B. combining a set of three p-values
   C. testing for association in a contingency table
   D. testing the hypothesis that the variance is equal to a specific value
   E. testing the hypothesis that two variances are equal

2. Which of the following is not true of a two-way contingency table?
   A. The row variable may be a dependent variable.
   B. The column variable may be a dependent variable.
   C. Both row and column variables may be dependent variables.
   D. Exactly one of the variables may be an independent variable.
   E. Neither the row nor the column variable may be controlled by the investigator.

3. Blood samples were taken from a sample of 100 medical students and serum cholesterol levels determined. A histogram suggested the serum cholesterol levels are approximately normally distributed. A chi-square goodness-of-fit test for normality yielded $\chi^2 = 9.05$ with 12 d.f. ($p = 0.70$). An appropriate conclusion is
   A. the data are consistent with the hypothesis that their distribution is normal
   B. the histogram is misleading in situations like this; a Poisson distribution would be more appropriate
   C. the goodness-of-fit test cannot be used for testing normality
   D. a scatter diagram should have been used to formulate the hypothesis
   E. none of the above

4. Two drugs—an active compound and a placebo—were compared for their ability to relieve anxiety. Sixty patients were randomly assigned to one or the other of the two treatments. After 30 days on treatment, the patients were evaluated in terms of improvement or no improvement. The study was double blind. A chi-square test was performed to compare the proportions of improved patients, resulting in $\chi^2 = 7.91$ with 1 d.f. ($p = 0.005$). A larger proportion improved in the active compound group. An appropriate conclusion is

A. the placebo group was handicapped by the random assignment to groups
B. an F-test is needed to evaluate the data
C. the data suggest the two treatments are approximately equally effective in relieving anxiety
D. the data suggest the active compound is more effective than placebo in relieving anxiety
E. none of the above

5. An investigator is studying the response to treatment for angina. Patients were randomly assigned to one of two treatments, and each patient's response was recorded in one of four categories. An appropriate test for the hypothesis of equal response patterns for the two treatments is the
   A. t-test
   B. F-test
   C. z-test
   D. χ^2-test
   E. rank sum test

6. For Problem 5, the appropriate number of degrees of freedom is
   A. 1
   B. 2
   C. 3
   D. 4
   E. 5

7. An investigator is studying the association between dietary and exercise habits in a group of 300 students. He summarizes the findings as follows:

Dietary Habits	Exercise Habits	Number Observed (O)	Number Expected (E)	O − E	Contribution to χ^2
Poor	Poor	23	27.45	−4.45	0.72
	Moderate	81	68.85	12.15	2.14
	Good	31	38.70	−7.70	1.53
Moderate	Poor	15	17.08	−2.08	0.25
	Moderate	47	42.84	4.16	0.40
	Good	22	24.08	−2.08	0.18
Good	Poor	23	16.47	6.53	2.59
	Moderate	25	41.31	−16.31	6.44
	Good	33	23.22	9.78	4.12
Total		300	300.00		$\chi^2 = 18.37$

The chi-square test with 4 d.f. was reported to be statistically significant ($p = 0.002$). Which of the following is true?

A. The correct number of degrees of freedom is 6.
B. The correct number of degrees of freedom is 8.
C. The chi-square is smaller than expected if there is no association.
D. The data are inconsistent with the hypothesis of no association.
E. The observed numbers tend to agree with those expected.

8. Data to be analyzed are arranged in a contingency table with 4 rows and 2 columns. The rows are four categories of an independent variable and the columns are a binomial response variable. The hypothesis of interest is that the proportion in the first column is the same for all categories of the independent variable. An appropriate distribution for the test statistic is
   A. Poisson
   B. standardized normal
   C. Student's t with 7 degrees of freedom
   D. F with 2 and 4 degrees of freedom
   E. chi-square with 3 degrees of freedom

9. A group of 180 students were interviewed to see how many follow a prudent diet. They were then given a 90-day series of in-depth lectures, including clinical evaluations on nutrition and its association with heart disease and cancer. One year later the students were re-interviewed and assessed for the type of diet they followed, yielding the following data:

Prudent Diet Initially	Prudent Diet at Follow-Up: Yes	No	Total
Yes	21	17	38
No	37	105	142
Total	58	122	180

McNemar's test results in $\chi^2 = 7.41$ with 1 d.f. ($p = 0.004$). An appropriate conclusion is
   A. the study is invalid since randomization was not used
   B. the effect of the lectures is confounded with that of the initial weight of the students
   C. the data suggest the lectures were ineffective
   D. the lectures appear to have had an effect
   E. none of the above

10. A researcher wishes to analyze data arranged in a 2 × 2 table in which each subject is classified with respect to each of two binomial variables. Specifically,

the question of interest is whether the two variables follow the same binomial distribution. A statistical test that is appropriate for the purpose is

A. McNemar's test
B. Wilcoxon's rank sum test
C. independent samples t-test
D. paired t-test
E. Mann-Whitney test

11. A lipid laboratory claimed it could determine serum cholesterol levels with a standard deviation less than 5 mg/dl. Samples of blood were taken from a series of patients. The blood was pooled, thoroughly mixed, and divided into aliquots. Ten of these aliquots were labeled with fictitious names and sent to the lipid laboratory for routine lipid analysis, interspersed with blood samples from other patients. Thus, the cholesterol determinations for these aliquots should have been identical, except for laboratory error. On examination of the data, the standard deviation of the 10 aliquots was found to be 7 mg/dl. Assuming cholesterol levels are approximately normally distributed, a chi-square test was performed of the null hypothesis that the standard deviation is 5; it was found that chi-square = 17.64 with 9 d.f. ($p = 0.04$). An appropriate conclusion is

A. the data are consistent with the laboratory's claim
B. the data suggest the laboratory's claim is not valid
C. rather than χ^2-test, a t-test is needed to evaluate the claim
D. the data fail to shed light on the validity of the claim
E. a clinical trial would be more appropriate for evaluating the claim

12. For which of the following purposes is the chi-square distribution not appropriate?

A. To test for association in a contingency table.
B. To construct a confidence interval for a variance.
C. To test the equality of two variances.
D. To test a hypothesis in large samples using the likelihood ratio criterion.
E. To combine p-values from independent tests of the same null hypothesis.

13. In a case-control study, the proportion of cases exposed to a suspected carcinogen is reported to be not significantly different from the proportion of controls exposed (chi-square with 1 d.f. = 1.33, $p = 0.25$). A 95% confidence interval for the odds ratio for these data is reported to be 2.8 ± 1.2. An appropriate conclusion is

A. there is no evidence that the suspected carcinogen is related to the risk of being a case
B. the reported results are inconsistent, and therefore no conclusion can be made

C. the p-value is such that the results should be declared statistically significant
D. we cannot study the effect of the suspected carcinogen in a case-control study
E. none of the above

14. An investigator performed an experiment to compare two treatments for a particular disease. He analyzed the results using a t-test and found $p = 0.08$. Since he had decided to declare the difference statistically significant only if $p < 0.05$, he decided his data were consistent with the null hypothesis. Several days later he discovered a paper on a similar previous study which reported $p = 0.11$. Further review of the literature produced two additional studies with p-values 0.19 and 0.07. Since the treatment differences were in the same direction in all four studies, the investigator computed

$$\begin{aligned} \chi^2 &= -2(\log_e 0.08 + \log_e 0.11 + \log_e 0.19 + \log_e 0.07) \\ &= 18.10 \text{ with 8 d.f. } (p = 0.013) \end{aligned}$$

An appropriate conclusion is
A. the investigator should not combine p-values from different studies
B. although none of the separate p-values is significant at the 0.05 level, the combined value is
C. the t-test should be used to combine p-values
D. the combined p-value is not statistically significant
E. the number of p-values combined is insufficient to warrant making a decision

15. The likelihood ratio is appealing because $-2 \log_e$(likelihood ratio) is known to be distributed as chi-square in large samples and this gives a criterion for
A. constructing a contingency table
B. determining the degrees of freedom in a t-test
C. narrowing a confidence interval
D. calculating the specificity of a test
E. evaluating the plausibility of a null hypothesis

CHAPTER NINE

Correlation and Regression

CHAPTER NINE

Key Concepts

linear regression:
- *slope*
- *intercept*
- *residual*
- *error sum of squares or residual sum of squares*
- *sum of squares due to regression*
- *mean squares*
- *error mean square*
- *regression (coefficient) of* x *on* y
- *least squares*

homoscedasticity, heteroscedasticity

linear relationship, covariance, product-moment correlation, rank correlation

multiple regression, stepwise regression, regression diagnostics, multiple correlation coefficient, partial correlation coefficient

regression toward the mean

Symbols and Abbreviations

b_0	sample intercept
$b_1, b_2, \ldots$	sample regression coefficients
MS	mean square
r	correlation coefficient
R	multiple correlation coefficient
R^2	proportion of the variability explained by a regression model
s_{XY}	sample covariance of X and Y (estimate)
SS	sum of squares
$\hat{y}$	value of y predicted from a regression equation
e	estimated error or residual from a regression model
β_0	population intercept
$\beta_1, \beta_2, \ldots$	population regression coefficients
ϵ	error or residual from a regression model (Greek letter epsilon)

Correlation and Regression

SIMPLE LINEAR REGRESSION

In Chapter 8 we discussed categorical data and introduced the notion of dependent and independent variables. We turn now to a discussion of relationships between dependent and independent variables of the continuous type. To explain such a relationship we often search for mathematical models such as equations for straight lines, parabolas, or other mathematical functions. The anthropologist Sir Francis Galton (1822–1911) used the term **regression** in explaining the relationship between the heights of fathers and their sons. From a group of 1078 father-son pairs he developed the following model, in which the heights are in inches:

$$\text{Son's height} = 33.73 + 0.516 \text{ (father's height)}$$

If we substitute 74 inches into this equation for the father's height, we arrive at about 72 inches for the son's height (i.e., the son is not as tall as his father). On the other hand, if we substitute 65 inches for the father's height, we find the son's height to be 67 inches (i.e., the son is taller than his father). Galton concluded that although tall fathers tend to have tall sons and short fathers tend to have short sons, the son's height tends to be closer to the average than his father's height. Galton called this "regression toward the mean." Although the techniques for modeling relationships among variables have taken on a much wider meaning, the term "regression" has become entrenched in the statistical literature to describe this modeling. Thus, nowadays we speak of regression models, regression equations, and regression analysis without wishing to imply that there is anything "regressive." As we shall see later, however, the phrase "regression toward the mean" is still used in a sense analogous to what Galton meant by it.

We begin by discussing a relatively simple relationship between two variables, namely a straight-line relationship. The equation $y = 3 + 2x$ is the equation of the straight line shown in Figure 9–1. The equation $y = 2 - 3x$ is shown in Figure 9–2, and the equation $y = 0.5x$ in Figure 9–3.

In general, the equation of a straight line can be expressed in the form

$$y = \beta_0 + \beta_1 x$$

where β_0 and β_1 are specified constants. Any point on this line has an x-coordinate and a y-coordinate. When $x = 0$, $y = \beta_0$; so the parameter β_0 is the value of the y-coordinate where the line crosses the y-axis and is called **the y-intercept.** In Figure 9–1, the intercept is 3, in Figure 9–2 it is 2. When $\beta_0 = 0$, the line goes through the origin (Figure 9–3) and the intercept is 0. The parameter β_1 is the **slope** of the line and measures the amount of change in y per unit increase in x. When β_1 is positive (Figs. 9–1 and 9–3), the slope is upward and x and y increase together; when β_1 is negative (Figure 9–2), the slope is downward and y decreases as x increases. When β_1 is zero, y is the same (it has value β_0) for all values of x and the line is horizontal (i.e., there is no slope). The parameter β_1 is undefined for vertical lines but approaches infinity as the line approaches a vertical position.

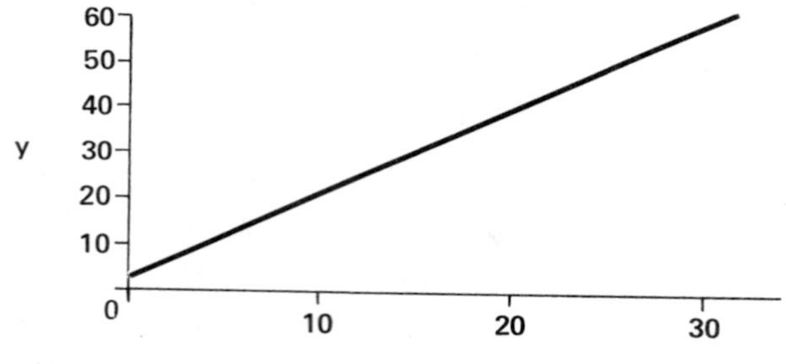

Figure 9–1. Graph of the equation $y = 3 + 2x$.

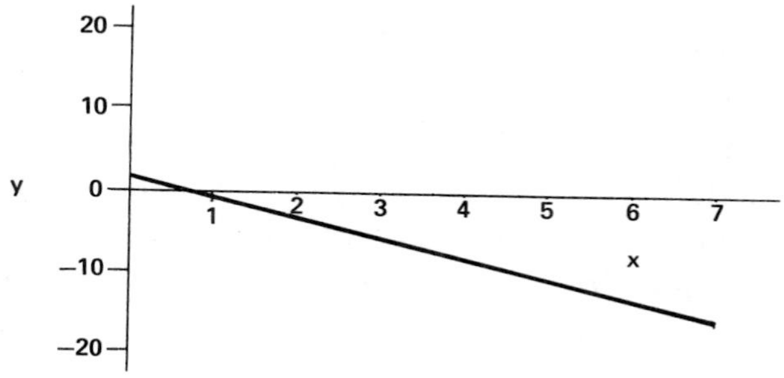

Figure 9–2. Graph of the equation $y = 2 - 3x$.

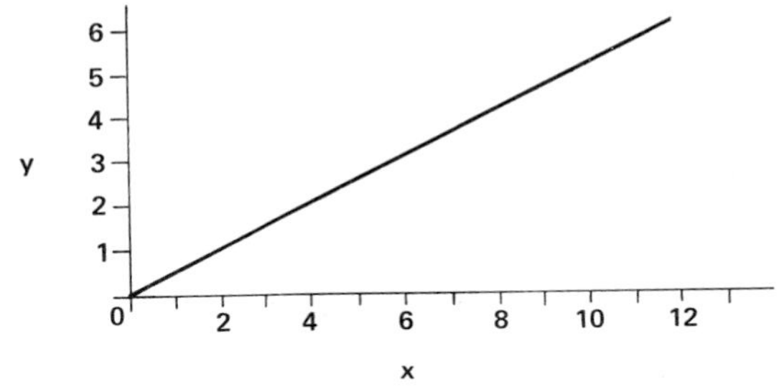

Figure 9–3. Graph of the equation y = 0.5x.

So far in our discussion we have assumed that the relationship between x and y is explained exactly by a straight line; if we are given x we can determine y—and vice versa—for all values of x and y. Now let us assume that the relationship between the two variables is not exact, because one of the variables is subject to random measurement errors. Let us call this random variable the dependent variable and denote it Y. The other variable x is assumed to be measured without error; it is under the control of the investigator and we call it the independent variable. This terminology is consistent with that of Chapter 8. In practice the independent variable will also often be subject to random variability caused by errors of measurement, but we assume that this variability is negligible relative to that of the dependent variable. For example, suppose an investigator is interested in the rate at which a metabolite is consumed or produced by an enzyme reaction. A reaction mixture is set up from which aliquots are withdrawn at various intervals of time and the concentration of the metabolite in each aliquot is determined. Whereas the time at which each aliquot is withdrawn can be accurately measured, the metabolite concentration, being determined by a rather complex assay procedure, is subject to appreciable measurement error. In this situation, time would be the independent variable under the investigator's control, and metabolite concentration would be the random dependent variable.

Since the relationship is not exact, we write

$$Y = \beta_0 + \beta_1 x + \epsilon$$

where ϵ, called the *error,* is the amount by which the random variable Y, for a given x, lies above the line (or below the line, if it is negative). This is illustrated in Figure 9–4. In a practical situation, we would have a sample of pairs of numbers, x_1 and y_1, x_2 and y_2, and so on. Then, assuming a straight line is an appropriate model, we would try to find the line that best fits the data. In other words, we would try to find estimates b_0 and b_1 of the param-

Figure 9–4. Graph of the regression equation y = 4 + 2x and errors for the points (x = 2, y = 4), and (x = 6, y = 17).

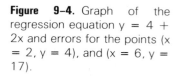

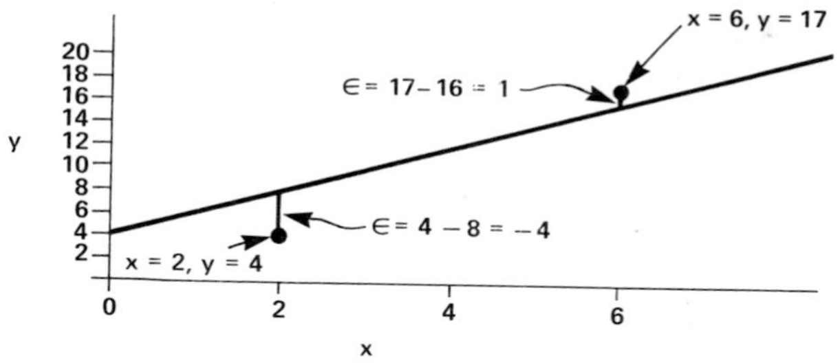

eters β_0 and β_1, respectively. One approach that yields estimates with good properties is to take the line that minimizes the sum of the squared errors (i.e., that makes the sum of the squared vertical deviations from the fitted line as small as possible). These are called **least-squares** estimates. Thus, if we have a sample of pairs, we can denote a particular pair (the ith pair) x_i, y_i, so that

$$y_i = \beta_0 + \beta_1 x_i + \epsilon_i$$

The ith error is

$$\epsilon_i = y_i - \beta_0 - \beta_1 x_i$$

as illustrated in Figure 9–4, and its square is

$$\epsilon_i^2 = (y_i - \beta_0 - \beta_1 x_i)^2$$

Then the least-squares estimates b_0 and b_1 of β_0 and β_1 are those estimates of β_0 and β_1, respectively, that minimize the sum of these squared deviations over all the sample values. The slope (β_1 or its least-squares estimate b_1) is also called the **regression of y on x**, or the **regression coefficient of y on x.**

Notice that if the line provides a perfect fit to the data (i.e., all the points fall on the line), then $\epsilon_i = 0$ for all i. Moreover, the poorer the fit, the greater the magnitudes, either positive or negative, of the ϵ_i. Now let us define the fitted line by $\hat{y}_1 = b_0 + b_1 x_i$, and the estimated error, or **residual**, by

$$e_i = y_i - \hat{y}_i = y_i - b_0 - b_1 x_i$$

Then a special property of the line fitted by least squares is that the sum of e_i over the whole sample is zero. If we sum the squared residuals e_i^2, we obtain a quantity called the **error sum of squares,** or **residual sum of squares.**

Figure 9–5. Graph of the regression equation $\hat{y}_i = b_0 + b_1x_i$ for the special case in which $b_1 = 0$ and $b_0 = \bar{y}$, with five residuals depicted.

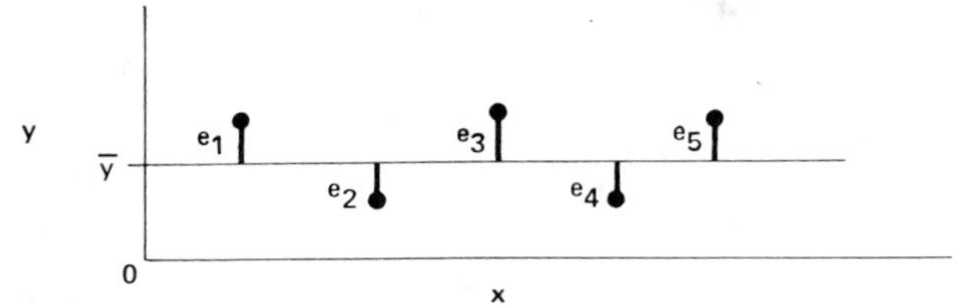

If the line is horizontal ($b_1 = 0$), as in Figure 9–5, the residual sum of squares is equal to the sum of squares about the sample mean. If the line is neither horizontal nor vertical, we have a situation such as that illustrated in Figure 9–6. The deviation of the ith observation from the sample mean, $y_i - \bar{y}$, has been partitioned into two components: (1) a deviation from the regression line, $y_i - \hat{y}_i = e_i$, the estimated error or residual; and (2) a deviation of the regression line from the mean, $\hat{y}_i - \bar{y}$, which we call the deviation due to regression (i.e., due to the straight-line model). If we square each of these three deviations ($y_i - \bar{y}$, $y_i - \hat{y}_i$, and $\hat{y}_i - \bar{y}$) and separately add them up over all the sample values, we obtain three sums of squares which satisfy the following relationship

Total sum of squared deviations from the mean
= sum of squared deviations from the regression model
+ sum of squared deviations due to, or
"explained by," the regression model.

We often abbreviate this relationship by writing

$$SS_T = SS_E + SS_R$$

where SS_T = total sum of squares about the mean
SS_E = error, or residual, sum of squares
SS_R = sum of squares due to regression

These three sums of squares have $n - 1$, $n - 2$, and 1 d.f., respectively. If we divide the last two sums of squares by their respective degrees of free-

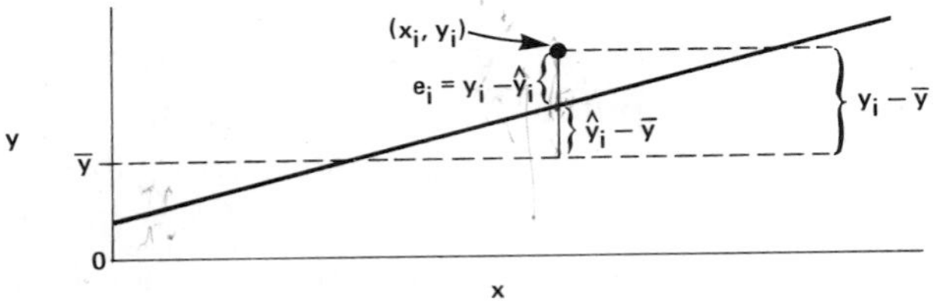

Figure 9–6. Graph of the regression equation $\hat{y}_i = b_0 + b_1x_i$, showing how the difference between each y_i and the mean $\bar{y}$ can be decomposed into a deviation from the line ($y_i - \hat{y}_i$) and a deviation of the line from the mean ($\hat{y}_i - \bar{y}$).

dom, we obtain quantities called **mean squares:** the **error,** or residual, **mean square** and the **mean square due to regression.** These mean squares are used to test for the significance of the regression, which in the case of a straight-line model is the same as testing whether the slope of the straight line is significantly different from zero. In the example discussed above, we may wish to test whether the metabolite concentration in the reaction mixture is in fact changing, or whether it is the same at all the different points in time. Thus we would test the null hypothesis H_0: $\beta_1 = 0$. Denote the error mean square MS_E and the mean square due to the straight line regression model MS_R. Then, under H_0 and certain conditions that we shall specify, the ratio

$$F = \frac{MS_R}{MS_E}$$

follows the F-distribution with 1 and $n - 2$ d.f. (*Note:* As is always true of an F-statistic, the first number of d.f. corresponds to the numerator, here MS_R, and the second to the denominator, here MS_E). As β_1 increases in magnitude, the numerator of the F-ratio will tend to have large values, and as β_1 approaches zero, it will tend toward zero. Thus, large values of F indicate departure from H_0, whether because β_1 is greater or less than zero. Thus if the observed value of F is greater than the 95th percentile of the F distribution, we reject H_0 at the 5% significance level for a two-sided test. Otherwise, there is no significant linear relationship between Y and x.

The conditions necessary for this test to be valid are the following:

1. For a fixed value of x, the corresponding Y must come from a normal distribution with mean $\beta_0 + \beta_1 x$.

2. The Y's must be independent.

3. The variance of Y must be the same at each value of x. This is called **homoscedasticity;** if the variance changes for different values of x, we have **heteroscedasticity.**

Furthermore, under these conditions, the least-squares estimates b_0 and b_1 are also the maximum likelihood estimates of $\beta_0 + \beta_1$.

The computation of b_0, b_1, and the mean squares, which nowadays are often automatically calculated on a computer or a pocket calculator, is detailed in Appendix 1. The statistical test is often summarized as shown in Table 9–1. Note that $SS_T/(n - 1)$ is the usual estimator of the (total) variance of Y if we ignore the x-values. The estimator of the variance of Y

TABLE 9–1. **Summary Results for Testing the Hypothesis of Zero Slope (Linear Regression Analysis)**

Source of Variability in Y	d.f.	Sum of Squares	Mean Square	F-ratio
Regression	1	SS_R	MS_R	MS_R/MS_E
Residual (error)	$n - 2$	SS_E	MS_E	
Total	$n - 1$	SS_T		

about the regression line, however, is MS_E, called the **error mean square** or the **mean squared error.** It estimates the error, or residual, variance not explained by the model.

Now b_1 is an estimate of β_1 and represents a particular value of an estimator which has a standard error that we shall denote s_{B_1}. Some computer programs and pocket calculators give these quantities instead of (or in addition to) the quantities in Table 9–1. Then, under the same conditions, we can test H_0: $\beta_1 = 0$ by using the statistic

$$t = \frac{b_1}{s_{B_1}}$$

which under H_0 comes from Student's t-distribution with $n - 2$ d.f. In fact, the square of this t is identical to the F-ratio defined earlier. So, in the case of simple linear regression, either an F-test or a t-test can be used to test the hypothesis that the slope of the regression line is equal to zero.

THE STRAIGHT-LINE RELATIONSHIP WHEN THERE IS INHERENT VARIABILITY

So far we have assumed that the only source of variability about the regression line is due to measurement error. But you will find that regression analysis is often used in the medical literature in situations in which it is known that the pairs of values x and y, even if measured without error, do not all lie on a line. The reason for this is not so much because such an analysis is appropriate, but rather because it is a relatively simple method of analysis, easily generalized to the case in which there are multiple independent variables (as we discuss later in this chapter), and readily available in many computer packages of statistical programs. For example, regression analysis might be used to study the relationship between triglyceride and cholesterol levels even though, however accurately we measure these variables, a large number of pairs of values will never fall on a straight line, but rather give rise to a scatter diagram similar to Figure 3–7. Regression analysis is not the appropriate statistical tool if, in this situation, we want to know how triglyceride levels and cholesterol levels are related in the population. It is, however, an appropriate tool if we wish to develop a prediction equation to estimate one from the other.

Let us call triglyceride level x and cholesterol level **Y**. Using the data illustrated in Figure 3–7, it can be calculated that the estimated regression equation of Y on x is

$$\hat{y} = 162.277 + 0.217x$$

and the residual variance is estimated to be 776 $(mg/dl)^2$. This variance includes both measurement error and natural variability, so it is better to call it "residual" variance rather than "error" variance. Thus, for a population of persons who all have a measured triglyceride level equal to x, we estimate that the random variable **Y**, measured cholesterol level, has mean 162.277 + 0.217x mg/dl and standard deviation $\sqrt{766} = 27.857$ mg/dl. This is the way we use the results of regression analysis to predict the distribution of cholesterol level (**Y**) for any particular value of triglyceride level (x). But we must not use the same equation to predict triglyceride level from cholesterol level. If we solve the equation

$$\hat{y} = 162.277 + 0.217x$$

for x, we obtain

$$x = -747.820 + 4.608\hat{y}$$

This is exactly the same line, the regression of **Y** on x, although expressed differently. It is not the regression of the random variable **X** on y. Such a regression can be estimated, and for these same data turns out to be

$$\hat{x} = -41.410 + 0.819y$$

with residual variance 2930 $(mg/dl)^2$. This is the equation that should be used to predict triglyceride level from cholesterol level. Figure 9–7 is a repeat of Figure 3–7 with these two regression lines superimposed on the data. The regression of **Y** on x is obtained by minimizing the sum of the squared **vertical** deviations (deviations in **Y**, i.e., parallel to the y-axis) from the straight line and can be used to predict the distribution of **Y** for a fixed value of x. The regression of **X** on y, on the other hand, is the converse

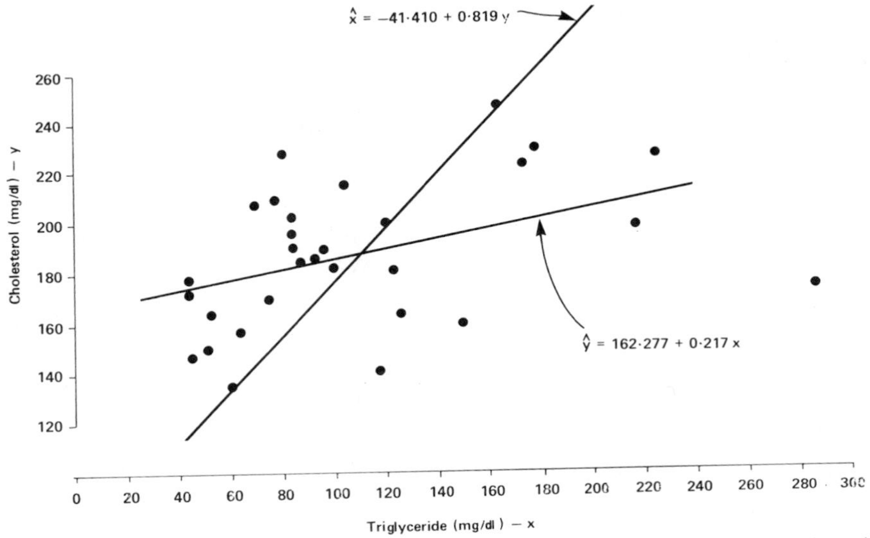

Figure 9–7. Scatterplot of serum cholesterol versus triglyceride levels of 30 medical students, with the two estimated regression lines.

situation: it is obtained by minimizing the sum of the squared *horizontal* deviations (deviations in the random variable X, i.e., parallel to the x-axis), and it is the appropriate regression to use if we wish to predict the distribution of X for a fixed (controlled) value of y. In this latter case, X is the dependent variable and y the independent variable.

It is clear that these are two different lines. Furthermore, the line that best describes the single, underlying **linear relationship** between the two variables falls somewhere between these two lines. It is beyond the scope of this book to discuss the various methods available for finding such a line, but you should be aware that such methods exist, and that they depend on knowing the relative accuracy with which the two variables X and Y are measured.

CORRELATION

In Chapter 3 we defined variance as a measure of dispersion. The definition applies to a single random variable. In this section we introduce a more general concept of variability called **covariance.** Let us suppose we have a situation in which two random variables are observed for each study unit in a sample, and we are interested in measuring the strength of the association between the two random variables in the population. For example, without trying to estimate the straight line relationship itself between cholesterol and triglyceride levels in male medical students, we might wish to estimate how closely the points in Figure 9–7 fit an underlying straight line. First we shall see how to estimate the covariance between the two variables.

Covariance is a measure of how two random variables vary together, either in a sample or in the population, when the values of the two random variables occur in pairs. To compute the covariance for a sample of values of two random variables, say X and Y, with sample means $\bar{x}$ and $\bar{y}$, respectively, the following steps are taken:

1. For each pair of values x_i and y_i, subtract $\bar{x}$ from x_i and $\bar{y}$ from y_i (i.e., compute the deviations $x_i - \bar{x}$ and $y_i - \bar{y}$).
2. Find the product of each pair of deviations [i.e., compute $(x_i - \bar{x})(y_i - \bar{y})$].
3. Sum these products over the whole sample.
4. Divide the sum of these products by one less than the number of pairs in the sample.

Suppose, for example, we wish to compute the sample covariance for X, Y from the following data:

i	x	y
1	10	30
2	20	50
3	30	70
4	40	90
5	50	110

Note that the sample means are $\bar{x} = 30$ and $\bar{y} = 70$. Following the steps outlined above, we

1. Subtract $\bar{x}$ from x_i and $\bar{y}$ from y_i

i	$x_i - \bar{x}$	$y_i - \bar{y}$
1	10 − 30 = −20	30 − 70 = −40
2	20 − 30 = −10	50 − 70 = −20
3	30 − 30 = 0	70 − 70 = 0
4	40 − 30 = 10	90 − 70 = 20
5	50 − 30 = 20	110 − 70 = 40

2. Find the products

i	$(x_i - \bar{x})(y_i - \bar{y})$
1	$(-20)(-40) = 800$
2	$(-10)(-20) = 200$
3	$(0)(0) = 0$
4	$(10)(20) = 200$
5	$(20)(40) = 800$

3. Sum the products: $800 + 200 + 200 + 800 = 2000$

4. Divide by one less than the number of pairs:

$$s_{XY} = \text{sample covariance} = \frac{2000}{5 - 1} = 500$$

Note in steps 1 and 2 that if both members of a pair are below their respective means (as in the case of the first pair, $i = 1$), the contribution to the covariance is positive (+800 for the first pair). It is similarly positive when both members of the pair are above their respective means (+ 200 and + 800 for $i = 4$ and 5, in the example). Thus, a positive covariance implies that X and Y tend to covary in such a manner that when one is either below or above its mean, so is the other. A negative covariance, on the other hand, would imply a tendency for one to be above its mean when the other is below its mean, and vice versa.

Now suppose X is measured in pounds and Y in inches. Then the covariance is in pound-inches, a mixture of units that is difficult to interpret. Recall that a variance is measured in squared units and we take the square root of the variance to get back to the original units. Obviously this does not work for the covariance. Instead, we divide the covariance by the product of the estimated standard deviation of X and the estimated standard deviation of Y, which we denote s_X and s_Y, respectively. The result is a pure, dimensionless number (no units) that is commonly denoted r and called the **correlation coefficient,** or **Pearson's product-moment correlation coefficient,** that is,

$$r = \frac{s_{XY}}{s_X s_Y}$$

where s_{XY} = sample covariance of X and Y
s_X = sample standard deviation of X
s_Y = sample standard deviation of Y

Thus, in the above example, $s_X = \sqrt{250}$ and $s_Y = \sqrt{1000}$, so

$$r = \frac{500}{\sqrt{250}\sqrt{1000}} = 1$$

In this example the correlation coefficient is +1. A scatterplot of the data indicates that all the points (x_i, y_i) lie on a straight line with positive slope, as illustrated in Figure 9–8(*a*).

If all the points lie on a straight line with negative slope, as in Figure 9–8(*b*), the correlation coefficient is −1. These are the most extreme values possible: a correlation can only take on values between −1 and +1. Figures 9–8(*a–h*) illustrate a variety of possibilities, and it can be seen that the magnitude of the correlation measures how close the points are to a straight line. Remember that the correlation coefficient is a dimensionless number, and so does not depend on the units of measurement. In Figures 9–8(*a–h*), the scales have been chosen so that the numerical value of the sample variance of Y is about the same as that of X—you can see that in each figure the range of Y is the same as the range of X. Now look at Figures 9–8(*i*) and (*j*). In each case the points appear to be close to a straight line, and you might therefore think that the correlation coefficient should be large in magnitude. If the scales are changed to make the range of Y the same as the range of X, however, Figure 9–8(*i*) becomes identical to Figure 9–8(*g*), and Figure 9–8(*j*) becomes identical to Figure 9–8(*h*). Once the scales have been adjusted, it becomes obvious that the correlation coefficient is near zero in each of these two situations. Of course, if all the points are on a horizontal line or on a vertical line, it is impossible to adjust the scales so that the range is numerically the same for both variables. In such situations, as illustrated in Figures 9–8(*k*) and (*l*), the correlation coefficient is undefined.

Note that the denominator in the correlation coefficient is the product of the sample standard deviations, which include both natural variability and measurement errors. Thus (unless the measurement errors in the two variables are themselves correlated), larger measurement errors automatically decrease the correlation coefficient. A small correlation between two variables can thus be due either to (1) little linear association between the two variables, or (2) large errors in their measurement. A correlation close to +1 or −1, on the other hand, implies that the measurement errors must be small relative to the sample standard deviations, and that the data points all lie close to a straight line. In fact, there is a close connection between

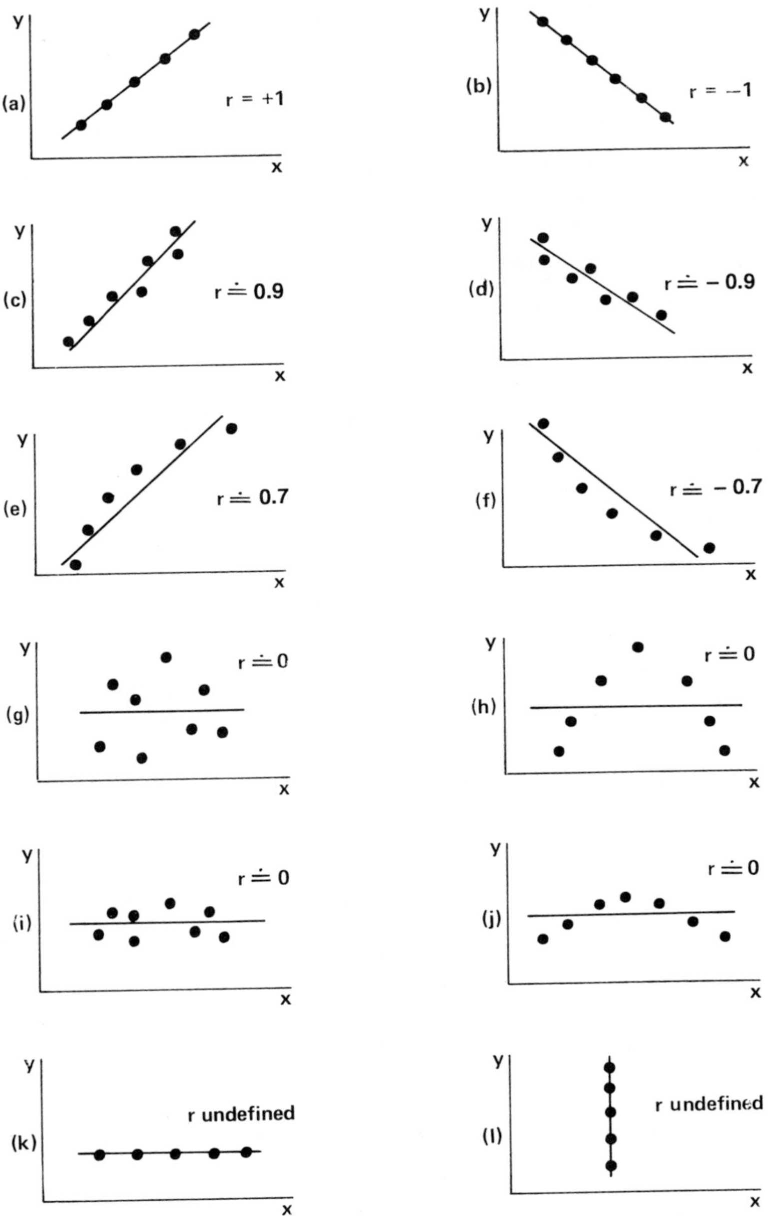

Figure 9–8. Scatterplots illustrating how the correlation coefficient, r, is a measure of the linear association between two variables.

the correlation coefficient and the estimated slope of the regression line. The estimated slope of the regression of Y on x is rs_Y/s_X, and the estimated slope of the regression of X on y is rs_X/s_Y. The correlation coefficient is significantly different from zero if, and only if, the regression coefficients are significantly different from zero; and such a finding implies a dependency between the two variables. However, a correlation coefficient of zero does not imply two variables are independent [see Figure 9–8(h)], and as we have seen before, a dependency between two variables does not necessarily imply a causal relationship between them.

SPEARMAN'S RANK CORRELATION

If we rank the x's from 1 to n (from largest to smallest, or vice versa), and we rank the y's from 1 to n in the same direction, and then compute the correlation coefficient between the pairs of ranks, we obtain the so-called **rank correlation coefficient** or **Spearman's rank correlation coefficient.** This correlation measures how closely the points can be fitted by a smooth, monotonic curve (i.e., a curve that is either always increasing or always decreasing). The rank correlations of the data in Figures 9–8(e) and (f) are $+1$, and -1, respectively. The curve that best fits the data in Figure 9–8(h), however, is not monotonic; it first increases and then decreases, and the rank correlation for these data is also approximately 0. Apart from being a measure of closeness to a monotonic curve, the rank correlation coefficient is less subject to influence by a few extreme values, and therefore sometimes gives a more reasonable index of association.

MULTIPLE REGRESSION

We have seen how a straight-line regression model can be fitted to data so that the variable x "explains" part of the variability in a random variable Y. A natural question to ask is, if one variable can account for part of the variability in Y, can more of the variability be explained by further variables? This leads us to consider models such as

$$Y = \beta_0 + \beta_1 x_1 + \beta_2 x_2 + \cdots \beta_q x_q + \epsilon,$$

where $x_1, x_2, \ldots, x_q$ are q distinct independent variables. Just as before, we can partition the total sum of squared deviations from the mean of y as follows:

$$SS_T = SS_E + SS_R,$$

where SS_E is the sum of squared deviations from the regression model and SS_R is the sum of squared deviations due to, or "explained by" the regression model. But now SS_E has $n - q - 1$ d.f. and SS_R has q d.f. Following the same line of reasoning as used in the case of simple linear regression, we can compute the quantities indicated in Table 9–2. Thus, $F = MS_R/MS_E$ with q and $n - q - 1$ d.f. and provides a simultaneous test of the hypothesis that all the regression coefficients are equal to zero; i.e.,

$$H_0: \beta_1 = \beta_2 = \cdots = \beta_q = 0$$

The sum of squares for regression can be further partitioned into q terms, each with 1 d.f., so that each coefficient can be tested separately. Thus, the results of a multiple regression analysis with three independent variables might look something like Table 9–3. The F-test provides an overall test of whether the coefficients β_1, β_2, and β_3 are simultaneously zero. The t-tests provide individual tests for each coefficient separately. These t-tests are to be interpreted only in light of the full model. Thus, if we drop x_2 from the model because the results suggest β_2 is not significantly different from zero, then we can fit a new model, say

$$Y = \beta_0' + \beta_1' x_1 + \beta_3' x_3 + \epsilon$$

to make inferences about the coefficients of x_1 and x_3 with x_2 removed from the model. Note in particular that β_1 in the full model is not equal to β_1' in the reduced model. Similarly,

$$\beta_3 \neq \beta_3'.$$

To study the effect of each independent variable fully, it is necessary to perform a regression analysis for every possible combination of the inde-

TABLE 9–2. **Summary Results for Multiple Regression**

Source of Variability in Y	d.f.	Sum of Squares	Mean Square	F
Regression	q	SS_R	MS_R	MS_R/MS_E
Residual (error)	$n - q - 1$	SS_E	MS_E	
Total	$n - 1$	SS_T		

TABLE 9–3. **Summary Results of Regression Analysis for the Model $Y = \beta_0 + \beta_1 x_1 + \beta_2 x_2 + \beta_3 x_3 + \epsilon$**

Source of Variability in Y	d.f.	Sum of Squares	Mean Square	F
Regression model	3	SS_R	MS_R	MS_R/MS_E
Error (residual)	$n - 4$	SS_E	MS_E	
Total	$n - 1$			

Parameter	Estimate	Standard Error of Estimate	t	p-Value
β_0	b_0	s_{B_0}	b_0/s_{B_0}	p_0
β_1	b_1	s_{B_1}	b_1/s_{B_1}	p_1
β_2	b_2	s_{B_2}	b_2/s_{B_2}	p_2
β_3	b_3	s_{B_3}	b_3/s_{B_3}	p_3

pendent variables. In the example of Table 9–3, this would entail conducting a regression analysis for each of the following models:

1. Y regressed on x_1
2. Y regressed on x_2
3. Y regressed on x_3
4. Y regressed on x_1 and x_2
5. Y regressed on x_1 and x_3
6. Y regressed on x_2 and x_3
7. Y regressed on x_1, x_2, and x_3

The larger the number of independent variables, the greater the number of possible combinations, and so it is often not feasible to perform all possible regression analyses. For this reason a stepwise approach is often used, though it may not lead to the best subset of variables to keep in the model. There are two types of **stepwise regression:** forward and backward.

1. *Forward stepwise regression* first puts into the model the single independent variable that explains most of the variability in Y, and then successively at each step inserts the variable that explains most of the remaining (residual) variability in Y. However, if at any step none of

the remaining independent variables explains a significant additional amount of variability in Y, at a predetermined level of significance, the procedure is terminated.

2. *Backward stepwise regression* includes all the independent variables in the model to begin with, and then successively at each step the variable that explains the least amount of variability in Y (in the presence of the other independent variables) is dropped from the model. However, a variable is dropped only if, at a predetermined level of significance, its contribution to the variability of Y (in the presence of the other variables) is not significant.

Whatever method is used to select among a set of independent variables in order to arrive at the "best" subset to be included in a regression model, it must always be remembered that the final result is merely a *prediction* model, and not necessarily a model for the *causation* of variability in the dependent variable.

Suppose a multiple regression analysis is performed assuming the model

$$Y = \beta_0 + \beta_1 x_1 + \beta_2 x_2 + \beta_3 x_3 + \epsilon$$

and, based on a sample of *n* study units, on each of which we have observations (y, x_1, x_2, x_3), we find

$$b_0 = 40 \qquad b_1 = 5 \qquad b_2 = 10 \qquad \text{and} \qquad b_3 = 7$$

as estimates of β_0, β_1, β_2, and β_3, respectively. The fitted regression model in this case is

$$\hat{y} = 40 + 5x_1 + 10x_2 + 7x_3$$

For each of the *n* study units we can substitute x_1, x_2, and x_3 into the fitted model to obtain a value $\hat{y}$. Suppose, for example, the observations on one of the study units were $(y, x_1, x_2, x_3) = (99, 4, 2, 3)$. On substituting the x's into the fitted model, we obtain

$$\hat{y} = 40 + 5(4) + 10(2) + 7(3) = 101$$

This procedure provides us with an estimate of the expected value of Y for the observed set of x's. We actually observed $y = 99$ for these x's; however, if we had observed a second value of Y for these same x's, that value would likely be some number other than 99. For each set of x's, our model assumes there is a distribution of y's corresponding to the random variable

Y. Thus, $\hat{y}$ is an estimate of the mean value of Y for that set of x's, and y $- \hat{y}$ (99 $-$ 101 $=$ 2, in our example) is the residual.

If we compute $\hat{y}$ for each of the *n* sample study units, and then compute the *n* residuals $y - \hat{y}$, we can examine these residuals to investigate the adequacy of the model. In particular, we can obtain clues as to whether

1. The regression function is linear
2. The residuals have constant variance
3. The residuals are normally distributed
4. The residuals are not independent
5. The model fits all but a few observations
6. One or more independent variables have been omitted from the model.

Methods for investigating model adequacy are called **regression diagnostics.** Regression diagnostics play an important role in statistical modeling because it is so easy to fit models with existing computer programs, whether or not those models are really appropriate. Use of good regression diagnostics will guard against blindly accepting misleading models.

Before leaving multiple regression, we should note that a special case involves fitting polynomial (curvilinear) models to data. We may have measured only one independent variable x, but powers of x are also included in the regression model as separate independent variables. For example, we may fit such models as

$$Y = \beta_0' + \beta_1' x + \beta_2' x^2 + \epsilon \qquad \text{(a quadratic model, or parabola)}$$

and

$$Y = \beta_0 + \beta_1 x + \beta_2 x^2 + \beta_3 x^3 + \epsilon \qquad \text{(a cubic model)}$$

MULTIPLE CORRELATION AND PARTIAL CORRELATION

Each sample observation y_i of the dependent variable corresponds to a fitted, or predicted, value $\hat{y}_i$ from the regression equation. Let us consider the pairs $(y_i, \hat{y}_i)$ as a set of data, and compute the correlation coefficient for these data. The result, called the **multiple correlation coefficient,** is denoted R. It is a measure of the overall linear association between the dependent variable Y and the set of independent variables $x_1, x_2, \ldots, x_q$ in the regression equation. In the special case that $q = 1$ (i.e., if there is only one independent variable), the multiple correlation coefficient R is simply equal to r, the correlation coefficient between X and Y. In general, R^2 equals the ratio SS_R/SS_T, and so measures the proportion of the variability explained by the regression model. If we fit a model with three independent variables and find $R^2 = 0.46$, and then fit a model that includes an additional, fourth independent variable and find $R^2 = 0.72$, we would conclude that the fourth variable accounts for an additional 26% (0.72 − 0.46) of the variability in Y. The square of the multiple correlation coefficient, R^2, is often reported when regression analyses have been performed.

The **partial correlation coefficient** is a measure of the strength of linear association between two variables after controlling for one or more other variables. Suppose, for example, we are interested in the correlation between serum cholesterol and triglyceride values in a random sample of men aged 20 to 65. Now it is known that both cholesterol and triglyceride levels tend to rise with age, so the mere fact that the sample includes men from a wide age range would tend to cause the two variables to be correlated in the sample. If we wish to discount after controlling for age (i.e., determine that part of the correlation that is over and above the correlation induced by common age), we would calculate the partial correlation coefficient, "partialing out the effect of" the variable age. The square of the partial correlation between the cholesterol and tryglyceride levels would then be the proportion of the variability in cholesterol level that is accounted for by the addition of triglyceride to a regression model that already includes age as an independent variable. Similarly, it would also equal the proportion of the variability in triglyceride level that is accounted for by the addition of cholesterol to a regression model that already includes age as an independent variable.

REGRESSION TOWARD THE MEAN

We conclude this chapter with a concept that, although it includes in its name the word "regression" and is indeed related to the original idea behind regression, is distinct from modeling the distribution of a random variable in terms of one or more other variables. Consider the highest three triglyceride values among those listed in Table 3–1 for 30 medical students. They are (in mg/dl) 218, 225, and 287, with a mean of 243.3.

Suppose we were to draw aliquots of blood from these three students on several subsequent days; should we expect the mean of the subsequent values to be 243.3? Alternatively, if we took later samples from those students with the three lowest values (45, 46, and 49), should we expect the average to remain the same (46.7)? The answer is that we should not: we should expect the mean of the highest three to become smaller, and the mean of the lowest three to become larger, on subsequent determinations. This phenomenon is called **regression toward the mean** and occurs whenever we follow up a selected, as opposed to a complete, or random, sample. To understand why regression toward the mean occurs, think of each student's measured triglyceride value as being made up of two parts: a "true" value (i.e., the mean of many, many determinations made on that student) and a random deviation from that true value; this latter could be due to measurement error and/or inherent variability in triglyceride value from day to day. When we select the three students with the highest triglyceride values based on a single measurement on each student, we tend to pick three that happen to have their highest random deviations, so that the mean of these three measurements (243.3 in our example) is most probably an overestimate of the mean of the three students' "true" values. Subsequent measures on these three students are equally likely to have positive or negative random deviations, so the subsequent mean will be expected to be the mean of their "true" values, and therefore probably somewhat lower. Similarly, if we pick the lowest three students in a sample, these single measurements will usually be underestimates of their true values, because they were probably picked partly because they happened to have their lowest random deviations when they were selected. If we were to make subsequent observations on the whole sample of 30 students, however, or on a random sample of them, regression toward the mean would not be expected to occur.

It is important to distinguish between regression toward the mean and a treatment effect. If subjects with high cholesterol levels are given a potentially cholesterol lowering drug, their mean cholesterol level would be expected to decrease on follow-up—even if the drug is ineffective—because of regression toward the mean. This illustrates the importance of having a control group taking a placebo, with subjects randomly assigned to the two groups. Regression toward the mean is then expected to occur equally in both groups, so that the true treatment effect can be estimated by comparing the groups.

SUMMARY

1. In the equation of a straight line, $y = \beta_0 + \beta_1 x$, β_0 is the y-intercept and β_1 is the slope.

2. In simple linear regression analysis, it is assumed that one variable (Y), the *dependent variable,* is subject to random fluctuations, whereas the other variable (x), the *independent variable,* is under the investigator's control. Minimizing the sum of the squared deviations of n sample values of Y from a straight line leads to the least-squares estimates b_0 of β_0 and b_1 of β_1, and hence the prediction line $\hat{y} = b_0 + b_1 x$. The sample residuals about this line sum to zero.

3. The total sum of squared deviations from the sample mean $\bar{y}$ can be partitioned into two parts: that due to the regression model and that due to error, or the residual sum of squares. Dividing these by their respective degrees of freedom gives rise to mean squares.

4. Under certain conditions the estimates b_0 and b_1 are maximum likelihood estimates, and the ratio of the mean squares (that due to regression divided by that due to residual) can be compared to the F-distribution with 1 and $n - 2$ d.f. to test the hypothesis $\beta_1 = 0$. These conditions are (a) for a fixed x, Y must be normally distributed with mean $\beta_0 + \beta_1 x$; (b) the Y's must be independent; and (c) there must be **homoscedasticity**—the variance of Y must be the same at each value of x.

5. It is possible to determine a standard error for b_1 and, under the same conditions, b_1 divided by its standard error comes from a t distribution with $n - 2$ d.f. In this situation, the F-test and the t-test are equivalent tests of the null hypothesis $\beta_1 = 0$.

6. The regression of Y on x can be used to predict the distribution of Y for a given value x, and the regression of X on y can be used to predict the distribution of X for a given value y. The line that best describes the underlying linear relationship between X and Y is somewhere between these two lines.

7. Covariance is a measure of how two random variables vary together. When divided by the product of the variables' standard deviations, the

result is the (product-moment) correlation, a dimensionless number that measures the strength of the linear association between the two variables. If all the points lie on a straight line with positive slope, the correlation is $+1$; if they all lie on a straight line with negative slope, the correlation is -1. A nonzero correlation between two random variables does not necessarily imply a causal relationship between them. A correlation of 0 implies no straight-line association—but there may nevertheless be a curvilinear association.

8. The rank correlation (computed from the ranks of the observations) measures how closely the data points fit a monotonic curve. The rank correlation is not greatly influenced by a few outlying values.

9. Multiple regression is used to obtain a prediction equation for a dependent variable Y from a set of independent variables x_1, x_2, ... The significance of the independent variables can be jointly tested by an F-ratio, or singly tested by t-statistics. A stepwise analysis is often used to obtain the "best subset" of the x-variables with which to predict the distribution of Y, but theoretically we can only be sure of reaching the best subset by examining all possible subsets. The prediction equation obtained need not reflect any causal relationship between the dependent and independent variables.

10. Regression diagnostics are used to investigate the adequacy of a regression model in describing a set of data. An examination of the residuals from a model gives clues as to whether the regression model is adequate, including whether the residuals are approximately normally distributed with constant variance.

11. The square of the multiple correlation coefficient is a measure of the proportion of the variability in a dependent variable that can be accounted for by a set of independent variables. The partial correlation coefficient between two variables is a measure of their linear association after allowing for the variables that have been "partialed out."

12. Whenever we make subsequent measurements on study units that have been selected for follow-up because they were extreme with respect to the variable being measured, we can expect regression toward the mean (i.e., study units with high initial values will tend to have lower values later, and study units with low initial values will tend to have higher values later).

FURTHER READING

Kleinbaum, D.G., Kupper, L.L., and Muller, K.E. (1988) Applied Regression and Other Multivariable Methods, ed. 2. PWS—Kent Publishing Co. (This book covers many aspects of regression analysis, including computational formulas. It requires only a limited mathematical background to read.)

PROBLEMS

1. Suppose two variables under study are temperature in degrees fahrenheit (y) and temperature in degrees centigrade (x). The "regression line" for this situation is

$$y = \frac{9}{5}x + 32$$

Assuming there is no error in observing temperature, the correlation coefficient would be expected to be

A. $\frac{9}{5}$
B. $\frac{5}{9}$
C. -1
D. $+1$
E. 0

2. An investigator studies 50 pairs of unlike-sex twins and reports that the regression of girl's birth weight (y) on boy's birth weight (x) is given by the following equation (all weights in grams):

$$y = 1221 + 0.403x$$

One can conclude from this that

A. the mean weight of twin brothers of girls who weigh 1000 g is predicted to be 1624 g
B. the mean weight of twin sisters of boys who weigh 1000 g is predicted to be 1624 g
C. the sample mean weight of the girls is 1221 g
D. the sample mean weight of the boys is 1221 g
E. the sample correlation between girl's weight and boy's weight is 0.403

3. In a regression analysis, the residuals of a fitted model can be used to investigate all the following except

A. the model fits all but a few observations
B. the error terms are normally distributed

C. the regression function is linear
D. the robustness of the rank sum test
E. one or more independent variables have been omitted from the model

4. Which of the following plots might represent data for variables X and Y that have a correlation coefficient equal to 0.82?

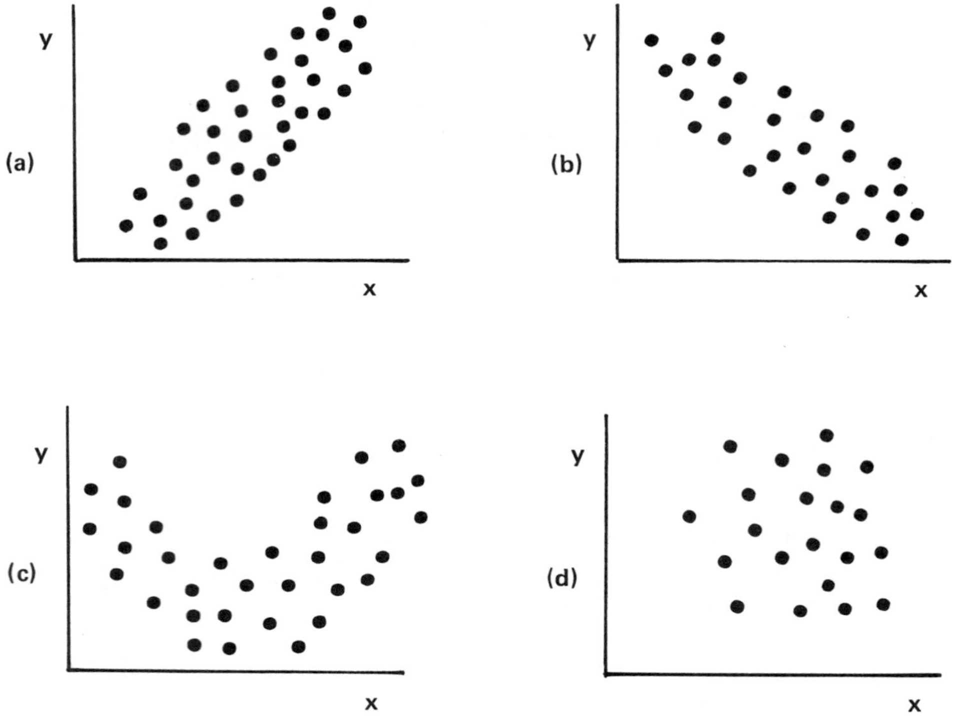

A. a
B. b
C. c
D. d
E. a and b

5. The correlation coefficient for the data in the graph below would be expected to have a value that is
A. a positive number of magnitude approximately equal to one
B. a negative number of magnitude approximately equal to one
C. a positive number of magnitude approximately equal to zero

D. a negative number of magnitude approximately equal to zero
E. none of the above

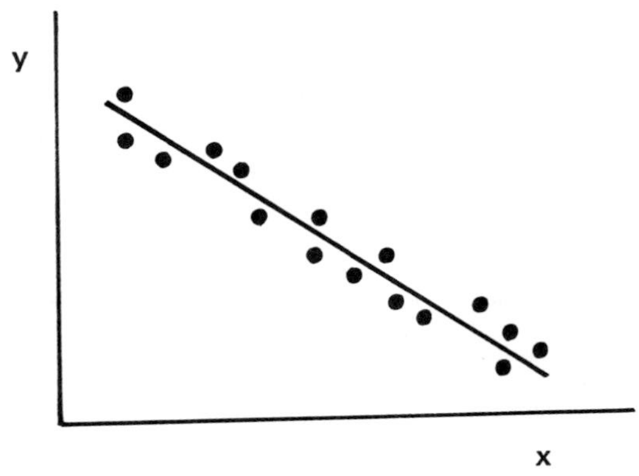

6. The Pearson correlation coefficient between variables A and B is known to be −0.50, and the correlation between B and C is known to be 0.50. What can we infer about the relationship between A and C?
   A. There is no association between A and C.
   B. A and C are independent.
   C. A and C are negatively correlated.
   D. A and C have a linear association.
   E. The relationship cannot be determined from the information given.

7. It is reported that both Pearson's product-moment correlation and Spearman's rank correlation between two variables are zero. This implies
   A. the two variables are independent
   B. the two variables tend to follow a monotonic curve
   C. there is no linear or monotonic association between the two variables
   D. all of the above
   E. none of the above

8. A data analyst is attempting to determine the adequacy of the model

$$y = \beta_0 + \beta_1 x_1 + \beta_2 x_2 + \epsilon$$

   for a particular set of data. The parameters of the model are
   A. least-squares estimates
   B. unbiased
   C. x_1 and x_2
   D. robust
   E. β_0, β_1, and β_2

9. A study was conducted to investigate the relationship between the estriol level of pregnant women and subsequent height of their children at birth. A scatter diagram of the data suggested a linear relationship. Pearson's product-moment correlation coefficient was computed and found to be r = 0.411. The researcher decided to re-express height in inches rather than centimeters and then recompute r. The recalculation should yield the following value of r:
   A. 0.000
   B. −0.411
   C. 0.411
   D. 0.500
   E. cannot be determined from data available

10. An investigator finds the following results for a regression analysis based on the model

$$Y = \beta_0 + \beta_1 x_1 + \beta_2 x_2 + \beta_3 x_3 + \epsilon$$

Source of Variability	d.f.	Sum of Squares	Mean Square	F	p-Value
Regression model	3	120	40	1.43	0.25
Error (residual)	36	1008	28		
Total	39	1128			

   Assuming all assumptions that were made in the analysis are justified, an appropriate conclusion is
   A. the mean of the dependent variable Y is not significantly different from zero
   B. the intercept term β_0 is significantly different from zero
   C. the parameters of the model are significantly different from zero
   D. the independent variables x_1, x_2, and x_3 do not account for a significant proportion of the variability in Y
   E. none of the above

11. A forward stepwise regression analysis was performed according to the following models:

Step 1: $Y = \alpha_0 + \alpha_1 x_1 + \epsilon$

Source of Variability	d.f.	Sum of Squares	Mean Square	F	p-Value
Regression model	1	135	135	13.5	0.001
Error (residual)	28	280	10		

Step 2: $Y = \beta_0 + \beta_1 x_1 + \beta_2 x_2 + \epsilon$

Source of Variability	d.f.	Sum of Squares	Mean Square	F	p-Value
Due to x_1	1	135	135		
Added to regression by x_2	1	62	62	7.68	0.01
Error (residual)	27	218	8.07		
Total	29	415			

The analysis was summarized as follows:

Step 1: $R^2 = 32.5\%$
Step 2: $R^2 = 47.5\%$

Assuming all assumptions made in the analysis are justified, an appropriate conclusion is

A. x_2 accounts for a significant amount of the variability in Y over and above that accounted for by x_1
B. neither x_1 nor x_2 accounts for a significant amount of the variability in Y
C. the proportion of variability explained by the regression model containing both x_1 and x_2 is less than should be expected in a stepwise regression analysis
D. the residual sum of squares is too large for meaningful interpretation of the regression analysis
E. the F-ratio is too small for interpretation in step 2

12. A multiple regression analysis was performed assuming the model

$$Y = \beta_0 + \beta_1 x_1 + \beta_2 x_2 + \beta_3 x_3 + \epsilon$$

The following results were obtained:

Parameter	Estimate	Standard Error of Estimate	t-Test	p-Value
β_0	40	14.25	2.81	0.005
β_1	5	2.43	2.06	0.025
β_2	10	38.46	0.26	0.600
β_3	7	2.51	2.79	0.005

Assuming all assumptions made in the analysis are justified, an appropriate conclusion is

A. none of the independent variables considered belong in the model
B. all of the independent variables considered belong in the model
C. x_1 and x_2 belong in the model, but x_3 does not

D. x_1 and x_3 belong in the model, but x_2 does not
E. x_2 and x_3 belong in the model, but x_1 does not

13. An investigator finds the following results for a regression analysis of data on 50 subjects based on the model

$$Y = \beta_0 + \beta_1 x_1 + \beta_2 x_2 + \beta_3 x_3 + \epsilon$$

Source of Variability	d.f.	Sum of Squares	Mean Square	F	p-Value
Regression model	3	360	120	4.29	<0.01
Error (residual)	46	1288	28		
Total	49	1648			

Assuming all assumptions that were made in the analysis are justified, an appropriate conclusion is
A. the mean of the dependent variable Y is not significantly different from zero
B. the intercept term β_0 is significantly different from zero
C. the parameters of the model are not significantly different from zero
D. the independent variables x_1, x_2, and x_3 account for a significant proportion of the variability in Y
E. none of the above

14. Regression diagnostics are useful in determining the
A. coefficient of variation
B. adequacy of a fitted model
C. degrees of freedom in a two-way contingency table
D. percentiles of the t-distribution
E. parameters of a normal distribution

15. A group of men were examined during a routine screening for elevated blood pressure. Those men with the highest blood pressure—namely those with diastolic blood pressure higher than the 80th percentile for the group—were reexamined at a follow-up examination 2 weeks later. It was found that the mean for the re-examined men had decreased by 8 mm Hg at the follow-up examination. The most likely explanation for most of the decrease is
A. the men were more relaxed for the second examination
B. some of the men became concerned and sought treatment for their blood pressure
C. observer bias
D. the observers were better trained for the second examination
E. regression toward the mean

CHAPTER TEN

Analysis of Variance and Linear Models

CHAPTER TEN

Key Concepts

analysis of variance (ANOVA):
  one-way or among-and-within groups
  nested or hierarchical
  two-way
  linear model
  fixed, random, and mixed models
  factorial arrangement of treatments

variance components

sum of squares, mean square, expected mean square

sampling fraction, fixed effects, random effects

simple effects, main effects, interaction effects, additive effects

analysis of covariance, covariate, concomitant variable

data transformation

Symbols and Abbreviations

F	percentile of the F distribution or corresponding test statistic
SS_A	sum of squares for factor A
MS_A	mean square for factor A
σ_A^2	variance component for factor A
SS_R	residual sum of squares
MS_R	residual mean square
σ_R^2	residual variance component

Analysis of Variance and Linear Models

Methods for comparing the means of two groups were discussed in Chapter 7. We now discuss the comparison of means when there are three or more groups. As an illustration, suppose we wish to investigate the mean effects of three treatments A, B, and C. More specifically, we wish to use sample data to test the null hypothesis that the three population means are identical, i.e., we wish to test

$$H_0: \mu_A = \mu_B = \mu_C$$

One approach to this problem is to perform all possible t-tests. In the present example, this involves three t-tests: one to test the hypothesis $\mu_A = \mu_B$, a second to test $\mu_A = \mu_C$, and a third to test $\mu_B = \mu_C$. There are a number of problems with this approach. In the first place, the tests are not independent; if $\mu_A = \mu_B$ and $\mu_A = \mu_C$, then it follows automatically that $\mu_B = \mu_C$. Thus, we can test any two of the three hypotheses and accomplish our goal. Which two of the three possible t-tests should we perform? In general, if there are a means to be compared, we have $a - 1$ independent pairs and hence $a - 1$ degrees of freedom in choosing such pairwise comparisons.

One of the assumptions of the t-test is that the variances are equal in the two groups being compared. If we have three or more groups, we must assume all pairs of groups have equal variances in order to perform the pairwise t-tests in the usual manner. This is equivalent to assuming that all the groups have the same variance, however, and if this is so, we can obtain a better estimate of the variance by pooling the information available in all the groups. Such a pooled estimate will have a greater number of degrees of freedom, and a test using this pooled estimate will have more power to detect differences among group means.

Even after eliminating the redundant pairwise comparisons and obtaining a pooled estimate of the variance, much effort will still be needed to compare all pairs of means with separate t-tests. It would be helpful if we had a single test to accomplish this goal. The F-test, which allows one to compare two variances, offers such a test. The strategy used is to compare the variability among the group means with the variability within the groups. If the variability among the sample group means is about what one might expect from the variability within the groups, we conclude that the group means are not significantly different. If, on the other hand, the variability among the sample group means is substantially larger than what we expect from the variability within the groups, we conclude that at least one of the pairwise comparisons is significant. Our criterion for evaluating the relative magnitude of these two sources of variability is the F-statistic. We compute a mean square deviation among group means and a pooled estimate of the mean square deviation within groups. Under the null hypothesis that the true population group means are equal, the ratio of these two mean squares—provided the observations are independent and come from normal distributions—follows an F distribution.

This approach to data analysis is an example of a general procedure, introduced by Sir Ronald Fisher, known as the analysis of variance. The **analysis of variance** (sometimes abbreviated as ANOVA) is essentially a procedure for partitioning the sum of squared deviations from the mean into components associated with recognized sources of variation. These components, which are sums of squares in their own right, are divided by their respective degrees of freedom to obtain mean squares. Ratios of mean squares are computed and compared with the F distribution to test hypotheses. Thus the analysis of variance is basically the same procedure as regression analysis. Whereas in regression analysis the independent variables are quantitative and usually continuous, however, in the analysis of variance the independent variables (such as "groups") are discrete categories.

If the F-test is used to compare more than two group means and it leads to the conclusion that the means are not all equal, then it is of interest to compare the means pairwise to discover which are significantly different. These comparisons should make use of the pooled estimate of the within-group variability. We must also take into consideration the increased probability, when performing multiple tests, of obtaining by chance at least one p-value that would lead to a conclusion that two means are significantly different. This is discussed further in Appendix 1.

In Chapter 2 we discussed a number of different experimental designs. For each design there is a corresponding analysis of variance. It is not possible in this brief chapter to cover the analysis corresponding to every design we have mentioned. Instead, we discuss the analysis of variance for a few special situations that illustrate the main principles involved.

COMPLETELY RANDOMIZED DESIGN WITH A SINGLE CLASSIFICATION OF TREATMENT GROUPS

First we consider the situation in which we have *a* treatment groups with the same number *n* of observations in each group. Thus, a total of *na* experimental units have been randomly assigned, *n* to each of the *a* treatments. The results of any analysis of variance can be summarized in a table similar to the tables we have already seen for regression analysis. Table 10–1 shows the table appropriate for this situation. As before, we use the abbreviations SS for sum of squares and MS for mean square. Formulas for calculating the sums of squares are given in Appendix 1. Once the sums of squares have been calculated, the corresponding mean squares are obtained by dividing by the appropriate number of degrees of freedom. Thus, in Table 10–1, $MS_A = SS_A/(a - 1)$, and $MS_R = SS_R/a(n - 1)$. Finally, the F-ratio is computed. As indicated in Table 10–1, the F-ratio to test the significance of the group means is MS_A/MS_R. This particular analysis is sometimes called a **one-way analysis of variance,** or an **among and within groups analysis of variance.**

Suppose, for example, an experiment is conducted in which 40 patients are randomly assigned to four treatment groups, so that 10 patients each receive one of the treatments A, B, C, and D, where the treat-

TABLE 10–1. **Outline of Analysis of Variance for Comparing *a* Group Means with *n* Observations per Group**

Source of Variability	Degrees of Freedom	Sum of Squares	Mean Square	F
Among groups	$a - 1$	SS_A	MS_A	MS_A/MS_R
Within groups (residual)	$a(n - 1)$	SS_R	MS_R	
Total	$an - 1$	SS_T		

TABLE 10–2. **Summary Statistics for Diastolic Blood Pressure in mm Hg after 4 Weeks of Treatment**

	Treatment			
	A	B	C	D
Number of patients	10	10	10	10
Mean	80	94	92	90
Standard deviation	10.5	9.5	9.7	10.2
Standard error of the mean	3.3	3.0	3.1	3.2

ments are various drugs aimed at lowering blood pressure. Suppose further that after 4 weeks of treatment the results shown in Table 10–2 are obtained. Clearly the mean diastolic blood pressure is lowest for treatment group A. But the question remains whether any of the differences are statistically significant. Since the four sample standard deviations are approximately equal, and diastolic blood pressure is approximately normally distributed, an F-test is appropriate for answering this question. The analysis of variance is given in Table 10–3. The F-ratio is 3.88, with 3 and 36 d.f. From Table 3, Appendix 2, we see that the 97.5th percentile of F with 3 and 36 d.f. lies between 3.59 and 3.46, and the 99th percentile lies between 4.51 and 4.31. Thus $0.01 < p < 0.025$ and we conclude that, at the 5% significance level, at least two of the means are different. It is clear from inspecting the mean blood pressures in Table 10–2 that the effect of treatment A is different from those of treatments B, C, and D, which are very much alike. Further statistical tests can be used (see Appendix 1) to determine which sets of means are, and which are not, significantly different.

These concepts are sometimes made clearer by introducing a mathematical model for the data. Specifically, we introduce a **linear model** as follows: Let any observation be denoted y_{ik}, where i indicates the group in

TABLE 10–3. **Analysis of Variance Corresponding to the Summary Results in Table 10–2**

Source of Variability	Degrees of Freedom	Sum of Squares	Mean Square	F
Among drug groups	3	1,160.00	386.67	3.88
Within drug groups	36	3,587.67	99.66	
Total	39	4,747.67		

which the observation belongs (i = 1, 2, . . . , a), and k denotes a specific observation in the ith group (k = 1, 2, . . . , n). We take as our linear model

$$y_{ik} = \mu_i + \epsilon_{ik}$$

where μ_i is the mean of the ith group and ϵ_{ik} is the random error, or residual, associated with the kth observation in the ith group. From this model we see that each observation is made up of two components, or effects: a mean effect, depending upon the group it belongs to, and a residual effect. Now consider the μ_i and ϵ_{ij} to be random variables. Let σ_A^2 be the variance of the μ_i (i.e., the variance among the population means), and let σ_R^2 be the variance of the ϵ_{ik} (i.e., the population residual variance). These are called **variance components.** Then if we take repeated samples, the expected, or mean, value of MS_A is $\sigma_R^2 + n\sigma_A^2$; and the expected, or mean, value of MS_R is σ_R^2. Note that under the null hypothesis the μ_i are all equal (and hence $\sigma_A^2 = 0$), and so these two mean squares then have the same expected value. It is necessary for two mean squares to have the same expected value if their ratio is to be distributed as F. As we shall see in later examples, this gives us a method of choosing the appropriate ratio of mean squares when we wish to test hypotheses in more complicated situations.

DATA WITH MULTIPLE CLASSIFICATIONS

Let us now turn to the situation in which the data to be analyzed can be classified in more than one way. For instance, we may be able to classify the data both into the a groups $A_1, A_2, \ldots, A_a$ and also into the b groups $B_1, B_2, \ldots, B_b$. These are thus two "factors," A and B. The way to analyze the data then depends on the way the data are classified by these independent variables, or factors. Furthermore, when there are multiple classifications, the way we choose the treatments for investigation and the population to which we wish to make inferences also determine the appropriate test statistic for any particular hypothesis.

Let us assume that the a treatments $A_1, A_2, \ldots, A_a$ are a subset of the total N_A possible treatments. Thus we have a sample of a out of N_A treatments, and we say the sampling fraction for treatments is a/N_A. The expected value of a mean square is often a function of the sampling fraction. If we do not wish to extrapolate beyond the treatments being investigated, $a = N_A$, and so the sampling fraction is 1. In this case we say the treatment effects are fixed. Suppose, for example, we have $a = 5$ nurses who are taking blood pressures and we wish to test the null hypothesis that there is no difference among the nurses with respect to the blood pressures they observe. If we are interested in making inferences only about these five particular nurses, then $N_A = 5$ and the sampling fraction is 1. We say that the "nurse effects" are **fixed.** On the other hand, suppose we consider these five nurses to be a random sample of all N_A nurses in a hospital, and the question is whether there are significant differences among the nurses *in this hospital* with respect to the blood pressures they observe. Then N_A is a larger number and the sampling fraction is less than 1. In this case we say the treatment effects (i.e., the "nurse effects") are **random.** If we wish to consider the five nurses to be a random sample of all nurses, so that we are testing whether there are differences among nurses in general with respect to blood pressures observed, then the sampling fraction is virtually zero.

The factors under study in an investigation may all be associated with fixed effects, they may all be associated with random effects, or some may be associated with fixed effects and others with random effects, so that we have a mixture of fixed and random effects. The models corresponding to these situations are called **fixed models, random models,** and **mixed models,** respectively.

NESTED INDEPENDENT VARIABLES

Let us consider the situation in which the treatment categories of multiple factors follow a **nested,** or **hierarchical,** arrangement, such as might arise from a multistage cluster-sampling scheme. For example, suppose we have a sample of a units of blood, which we denote $A_1, A_2, \ldots, A_a$. From each unit we take b aliquots, and then on each aliquot we make n replicate observations. This arrangement of the observations is illustrated in Figure 10–1, and the corresponding **nested,** or **hierarchical, analysis of variance** is outlined in Table 10–4. The units of blood are the groups of type A and the aliquots are the groups of type B. Computational formulas for the sums of squares are given in Appendix 1, and the mean squares are obtained, as always, by dividing each sum of squares by its number of d.f. Note that within each group of observations of type A there are $b - 1$ d.f. among the groups of type B; hence, the total number of d.f. among all the groups of type B, within the groups of type A (among aliquots within the a units, in our example), is $a(b - 1)$. Similarly, there are $n - 1$ d.f. among the rep-

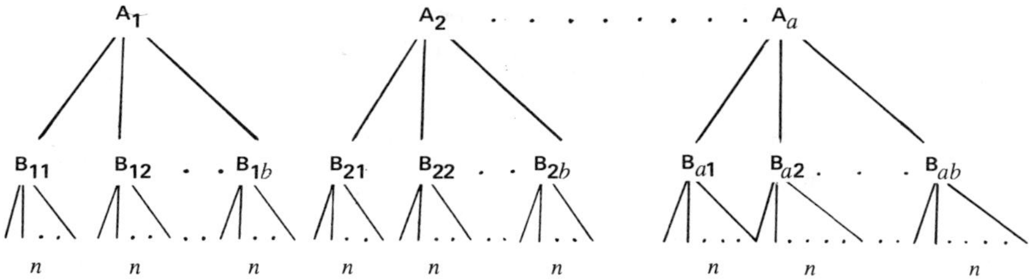

Figure 10–1. Nested arrangement of treatment categories: *n* replicate observations on each of *b* aliquots from each of *a* units of blood.

licate observations within any group of type B, and hence, as there are altogether ab groups of type B, there are $ab(n - 1)$ d.f. among replicates within groups of type B. Finally, note that the total number of d.f. adds up to one less than the total number of observations, i.e.,

$$(a - 1) + a(b - 1) + ab(n - 1) = abn - 1$$

This equality must hold for any analysis of variance table.

It can be seen in Table 10–4 that the appropriate F-statistic to test whether there are significant differences among the groups of type B (among aliquots within units, in our example) is $MS_{B|A}/MS_R$. The appropriate statistic (F_A) to test whether there are significant differences among the groups of type A, however, depends on the sampling fraction b/N_B. This can be seen by inspecting the expected mean squares in Table 10–5, which are functions of three variance components: σ^2_A, the variance among the group effects of type A; $\sigma^2_{B|A}$, the variance of the group effects of type B within the groups of type A; and σ^2_R, the variance of the residual effects within groups of type B. First, note that when $\sigma^2_{B|A} = 0$, $MS_{B|A}$ and MS_R have the same expected values, confirming that the F-statistic $MS_{B|A}/MS_R$ is appropriate to test for significant differences among the groups of type

TABLE 10–4. **Outline of Analysis of Variance for Two Nested Factors with *n* Replicate Observations per Treatment**

Source of Variability	Degrees of Freedom	Sum of Squares	Mean Square	F
Among groups of type A	$a - 1$	SS_A	MS_A	F_A
Among groups of type B within groups of type A	$a(b - 1)$	$SS_{B\|A}$	$MS_{B\|A}$	$MS_{B\|A}/MS_R$
Among replicates within groups of type B (residual)	$ab(n - 1)$	SS_R	MS_R	
Total	$abn - 1$	SS_T		

Table 10-5. **Expected Values of the Mean Squares for the Analysis of Variance in Table 10-4**

Mean Square	Expected Mean Square
MS_A	$\sigma_R^2 + \left(1 - \frac{b}{N_B}\right) n\sigma_{B\mid A}^2 + bn\sigma_A^2$
$MS_{B\mid A}$	$\sigma_R^2 + n\sigma_{B\mid A}^2$
MS_R	σ_R^2

B, regardless of the sampling fraction b/N_B. To test for significant differences among the groups of type A, we need to divide MS_A by a mean square that has the same expected value when $\sigma_A^2 = 0$ [i.e., by a mean square that has expected value $\sigma_R^2 + (1 - b/N_B)n\sigma_{B\mid A}^2$]. If $b = N_B$, then $1 - b/N_B = 0$ and $F_A = MS_A/MS_R$ is appropriate. If, on the other hand, N_B is very large compared with b, then $b/N_B \doteq 0$ and $F_A = MS_A/MS_{B\mid A}$ is appropriate. Consider our example in which b aliquots are taken from each unit of blood. A unit of blood comprises 500 ml. If we divide each unit into five 100-ml aliquots, so that $b = N_B = 5$, the aliquot effects are fixed and the divisor in F_A should be MS_R. If, on the other hand, we take a sample of five 0.1-ml aliquots from the total of 5,000 such aliquots (500 ml ÷ 0.1 = 5,000), so that $b = 5$ and $N_B = 5{,}000$, then the aliquot effects are random; and because in this instance $b/N_B \doteq 0$, the divisor in F_A should be $MS_{B\mid A}$. If b/N_B is neither unity nor close to zero, then we must use an approximate test, but this is beyond the scope of this book.

This analysis of variance can be extended to any number of nested factors, and it is not necessary to have the same number of replicates within each treatment of a particular type, or the same number of treatment categories of a given type within each category of another type. There are computer programs that produce the analysis of variance table and calculate F-statistics. Many of these programs, however, use MS_R as the divisor for all the F-tests, regardless of whether or not it is appropriate. In other words, many computer programs calculate F-statistics that are correct only if the effects of all the independent variables are fixed.

CROSS-CLASSIFIED INDEPENDENT VARIABLES

In Chapter 2 we described the **factorial arrangement** of treatments, in which the treatments comprise all possible combinations of different levels of two or more factors. Thus if each factor is a drug and each level a different dose, then in a factorial arrangement the treatments comprise all possible combinations of dose levels of the drugs. Suppose, for example, we study two different dose levels of drug A in combination with three different dose levels of drug B, so that altogether there are six distinct treatments: A_1B_1, A_1B_2, A_1B_3, A_2B_1, A_2B_2, and A_2B_3. (In general we could have a levels of factor A and b levels of factor B, and so a total of ab treatments.) The resulting data are then *cross-classified.* The mean responses to the treatments can also be cross-classified, as the following two-way table shows:

		Factor B		
		B_1	B_2	B_3
Factor A	A_1	μ_{11}	μ_{12}	μ_{13}
	A_2	μ_{21}	μ_{22}	μ_{23}

Thus, μ_{11} is the (population, or true) effect of the treatment A_1B_1, μ_{12} that of the treatment A_1B_2, and so forth. These means are called **simple effects.**

If in a study, n sample units are randomly assigned to each treatment, we can describe the data by the following linear model

$$y_{ijk} = \mu_{ij} + \epsilon_{ijk}$$

where $i = 1, 2$ (in general, $i = 1, 2, \ldots, a$)
$j = 1, 2, 3$ (in general $j = 1, 2, \ldots, b$)
$k = 1, 2, \ldots, n$

Thus, ϵ_{ijk} is a random amount by which the response of the kth replicate differs from μ_{ij}, the mean response (of all possible study units) to treatment A_iB_j. This model is similar to the model we introduced for analyzing data with a single classification factor, and in fact we can analyze the data as though there are just six different groups, testing the null hypothesis that the six group means (simple effects) are all equal.

When we have a factorial arrangement of treatments, however, it is often of interest to ask questions about each factor separately. We may ask, for example, whether there is any difference in mean response to A_1 and A_2, regardless of the level of B; or whether there are any significant differences among the mean responses to B_1, B_2, and B_3, regardless of the level of A. These mean responses are called **main effects.** Thus the main effect of treatment A_1 is the average of μ_{11}, μ_{12}, and μ_{13}, and the main effect of A_2 is the average of μ_{21}, μ_{22}, and μ_{23}. Similarly, the main effect of B_1 is the average of μ_{11} and μ_{21}, and so on. These averages are usually taken to be unweighted averages [e.g., $(\mu_{11} + \mu_{12} + \mu_{13})/3$ for the main effect of A_1]. Sometimes, however, the main effects are defined as weighted averages, some simple effects being given more weight than others in the averaging process. Except in a special situation that will now be described, the difference between the main effects of A_1 and A_2, and the differences among the main effects of B_1, B_2, and B_3, depend on how the main effects are defined. Suppose the differences between the simple effects of A_1 and A_2 are the same at all levels of B, i.e., suppose

$$\mu_{11} - \mu_{21} = \mu_{12} - \mu_{22} = \mu_{13} - \mu_{23}$$

Then it does not matter how the simple effects are weighted in the definition of the main effects—the difference between the main effect of A_1 and the main effect of A_2 is always the same, and the difference between the main effects of any two particular levels of B is always the same. Equivalently, the same result holds (i.e., the differences between main effects do not depend on how the simple effects are weighted in the definition of the main effects) if the level of A has no effect on the difference between simple effects of B_1, B_2, and B_3, i.e., if

$$\mu_{11} - \mu_{12} = \mu_{21} - \mu_{22}$$

and

$$\mu_{12} - \mu_{13} = \mu_{22} - \mu_{23}$$

In this situation we say there is no **interaction** between A and B, or that the effects of the different levels of A and B are **additive.** When the means are plotted on a graph and the simple effects making up each main effect are joined by straight lines, the lines are parallel, as in Figure 10–2. If the lines are not parallel, as in Figure 10–3, then an *interaction* is present and the different levels of A and B are not all additive. Thus, if $\mu_{11} - \mu_{21}$ is equal to $\mu_{12} - \mu_{22}$, which is the same as $\mu_{11} - \mu_{12}$ being equal to $\mu_{21} - \mu_{22}$, there is no interaction between the two levels of A and the first two levels of B; if these differences are not equal, then the difference between them, or

$$\mu_{11} - \mu_{21} - (\mu_{12} - \mu_{22})$$

is an interaction effect. Similarly

$$\mu_{12} - \mu_{13} - (\mu_{22} - \mu_{23})$$

is an interaction effect between the two levels of A and the second and third levels of B.

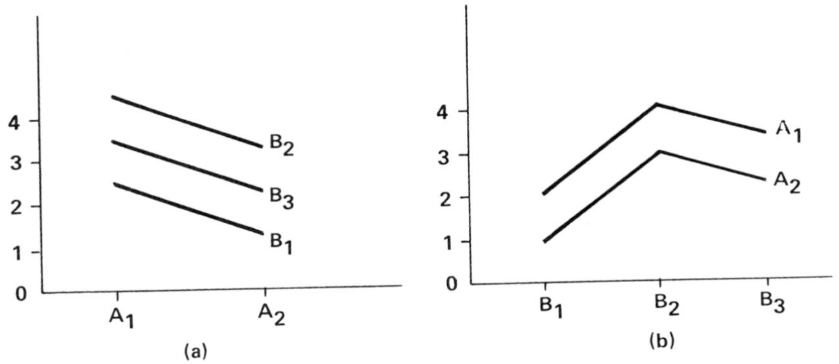

Figure 10–2. Mean responses in a case in which there is no interaction: (a) for different levels of factors B plotted against the levels of factor A; (b) for different levels of factor A plotted against the levels of factor B.

In practice, the presence or absence of interaction is obscured by chance variability in the data. We therefore test whether the interaction is significant (i.e., whether the observed departure of the sample means from additivity is too large to be explained by the chance variation within each treatment group). Similarly, we can test whether each of the main effects is significant (i.e., whether there are significant differences among the main effects of each factor). The analysis of variance table for a levels of A and b levels of B, with n replicates of each treatment, is given in Table 10–6. This is often called a **two-way analysis of variance.** The F-statistic to test for the presence of interaction is MS_{AB}/MS_R, but the appropriate divisors in the statistics F_A and F_B to test for differences among the main effects depend on whether the effects are fixed or random.

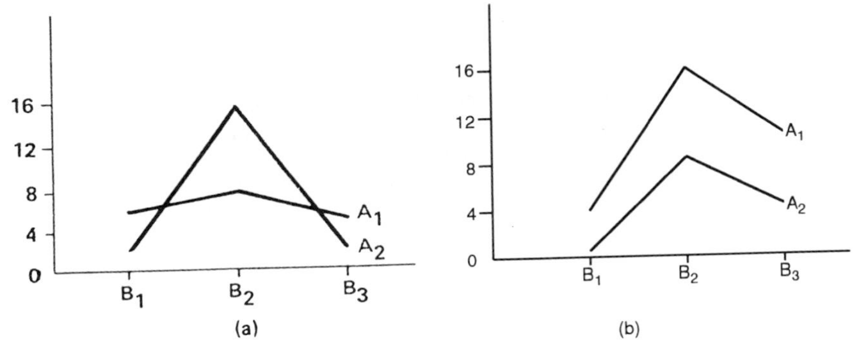

Figure 10–3. Mean responses in a case in which there is interaction, the different levels of factor A being plotted against the levels of factor B. In case (a), no transformation can eliminate the interaction; in case (b), the square-root transformation eliminates the interaction: plotting the square roots of the means results in Figure 10–2 (b).

Table 10–6. **Outline of Analysis of Variance for a Factorial Arrangement of Treatments: Two Factors and *n* Replicate Observations per Treatment**

Source of Variability	d.f.	Sum of Squares	Mean Square	F
Main effects of A	$a - 1$	SS_A	MS_A	F_A
Main effects of B	$b - 1$	SS_B	MS_B	F_B
A × B interaction	$(a - 1)(b - 1)$	SS_{AB}	MS_{AB}	MS_{AB}/MS_R
Residual	$ab(n - 1)$	SS_R	MS_R	
Total	$nab - 1$	SS_T		

Table 10–7 gives the expected values of the mean squares in terms of the following four variance components:

σ_A^2 the variance among the main effects of A

σ_B^2 the variance among the main effects of B

σ_{AB}^2 the variance among the interaction effects

σ_R^2 the residual variance within the *ab* treatments

Inspection of this table reveals that if the main effects of a factor are fixed (so that the sampling fraction is 1), the appropriate divisor for the F-statistic to test for significant differences among the main effects of the other factor is MS_R. If the main effects of a factor are random and the sampling fraction is near 0, however, the appropriate denominator for the other factor is MS_{AB}.

Suppose, for example, we are interested in comparing two different types of surgery. Eligible patients are randomly assigned to each type of surgery and to each of four different surgeons. The two different types of

Table 10–7. **Expected Values of the Mean Squares for the Analysis of Variance in Table 10–6**

Mean Square	Expected Mean Square
MS_A	$\sigma_R^2 + \left(1 - \frac{b}{N_B}\right) n\sigma_{AB}^2 + bn\sigma_A^2$
MS_B	$\sigma_R^2 + \left(1 - \frac{a}{N_A}\right) n\sigma_{AB}^2 + an\sigma_B^2$
MS_{AB}	$\sigma_R^2 + n\sigma_{AB}^2$
MS_R	σ_R^2

surgery are fixed effects because they can hardly be considered a random sample of types of surgery: we chose these two particular types for study. We might wish to consider the four surgeons as a random sample, however, if we are interested in comparing the two types of surgery when performed by all surgeons, not just the particular four in our study. (Of course, the four surgeons were most probably *not* a random sample of all surgeons; but if each is in a different location, and on the basis of several criteria they appear to be typical of the general population of surgeons, it could be of interest to analyze the results of the study as though the surgeons were a random sample.) In this situation, if we let the type of surgery be factor A, we have $a = N_A = 2$ with sampling fraction $a/N_A = 1$. Also, letting the different surgeons be factor B, we have $b = 4$ and N_B large, so that the sampling fraction b/N_B is near zero. Then, to test whether there is a significant difference between the two types of surgery, the appropriate divisor is MS_{AB}, whereas to test whether there are significant differences among surgeons the appropriate divisor is MS_R.

Notice that if we use MS_{AB} as the divisor when we should use MS_R, the F-statistic will tend on an average to be too small. The test will be *conservative,* tending too often not to reject the null hypothesis of no differences among the main effects. If, on the other hand, we use MS_R as the divisor when we should use MS_{AB}, the test will tend on an average to be *liberal*—the null hypothesis of no main effects will tend to be rejected too frequently. If there is no interaction ($\sigma^2_{AB} = 0$), however, both MS_{AB} and MS_R have the same expected value and either can be used as the divisor. In this situation the best procedure is to pool the two mean squares to obtain an estimate of σ^2_R with more d.f. The pooling is accomplished by adding together the corresponding sums of squares SS_{AB} and SS_R, and then dividing by the sum of their d.f.:

$$(b - 1)(b - 1) + ab(n - 1)$$

This pooled mean square, with $(a - 1)(b - 1) + ab(n - 1)$ d.f., is then used as the divisor in F-tests for each of the main effects.

If there is only one study unit for each treatment (i.e., there is no replication and $n = 1$), there are no residual d.f. and MS_R does not exist. Nevertheless, we can still test each of the main effects using MS_{AB}, with $(a - 1)(b - 1)$ d.f. as divisor. At worst, this test may be conservative (if $\sigma^2_{AB} \neq 0$ and the main effects of one of the factors are random), and so not too powerful. If, however, one of the main effects is judged to be significant at a particular level on the basis of such a test, then we can be sure that there is indeed significance at that level. If no interaction is present, it is easier to interpret a significant difference between two main effects. Sometimes it is possible to transform the measurements in such a manner that no interaction is present. For example, if we plot the square roots of the six means depicted in Figure 10–3(*b*), we obtain Figure 10–2(*b*). Thus there is no interaction among these means after transformation to the square-root scale. On the other hand, if there is a crossing of the lines as depicted in Figure 10–3(*a*), no transformation will eliminate the interaction. Biological systems are often governed by factors that act multiplicatively, rather than additively, in which case a logarithmic transformation will bring about additivity (i.e., remove interaction).

ANALYSIS OF COVARIANCE

In both regression analysis and the analysis of variance, we assume the data follow a linear model (i.e., that the dependent variable is a linear function of the independent variable[s] and a random error). In regression analysis, the independent variables are quantitative and usually continuous, whereas in the analysis of variance the independent variables are always discrete. If our linear model contains both types of independent variable, the corresponding analysis is called an **analysis of covariance.**

As an example, suppose we wish to compare three different drug treatments for their effects on blood pressure, but, because we know blood pressure changes with age, we wish to include age in our linear model. Thus we have two independent variables: the quantitative variable age and the discrete variable drug treatment. The analysis would follow the same principles we have discussed for regression analysis and the analysis of variance, but would be called an analysis of covariance. The quantitative trait age is called a covariate, and the purpose of the analysis is to determine whether, after allowing for the effects of age, the treatment effects are significantly different. In this type of analysis we assume that the effect of age, the covariate, is the same in all three drug treatment groups; that is, we assume that the regressions of blood pressure on age in the three groups are parallel lines. There are two reasons to allow for the effect of age. First, even though there is random assignment of patients to the different drug treatments, the three groups will probably still differ somewhat in their age distributions. An analysis of covariance compares the treatment effects "adjusted" to a common mean age for the three groups (i.e., as though the three groups had the same mean age). Second, even if the three groups have exactly the same mean age, allowing for age as a covariate identifies this extra source of variation and excludes it from the residual mean square. The residual mean square is thus smaller, and as a consequence, our tests of hypotheses are more powerful. Sources of variability that are randomly distributed among treatment groups do not affect the validity of significance tests, but do affect their power. An analysis that takes account of these sources, excluding from the residual mean square any variability due to them, is usually more powerful. **Concomitant variables,** which were discussed in Chapter 2, are for this reason often taken to be covariates in an analysis of covariance. The results of an analysis of covariance must be carefully scrutinized, however, because they can be misleading if the regression on a covariate (the slope of the straight line for predicting the dependent variable from that covariate) is not the same in all groups.

ASSUMPTIONS ASSOCIATED WITH THE ANALYSIS OF VARIANCE

The F-distribution is the theoretical model against which an F statistic is evaluated. To be valid, hypothesis-testing procedures using the F distribution in the analysis of variance require the following assumptions:

1. The numerator and denominator of the F-ratio are independent.
2. The observations represent random samples from the populations being compared.
3. The observations are drawn from normally distributed populations.
4. The variances of all the populations are equal.

If the data fail to satisfy these assumptions, then the stated significance levels may be in error. For example, the F-tables may suggest that the p-value is 0.05 when the true p-value is 0.03 or 0.06. Whereas this small error may not be important, it is the uncertainty of its magnitude that is disturbing. Since it may be impossible to be certain that all the assumptions are satisfied exactly, F-tests in the analysis of variance are often viewed as approximate rather than exact.

The first two assumptions are satisfied if there is randomization. Random allocation of study units to the comparison groups virtually assures that the numerator and denominator of the F-ratio are independent, and random selection of the study units allows one to make valid inferences to populations.

Unless departure from normality is so extreme that it can be readily detected by visual inspection of the data, lack of normality has little effect on the test. The analysis of variance F-test to compare means is fairly robust against nonnormality, whether in terms of skewness or kurtosis. This is in contrast to the F-test to compare two variances, discussed in Chapter 7, which is very sensitive to nonnormality. In the case of marked skewness, it may be more appropriate to compare population medians rather than means. In such instances, it may be possible to transform the data (e.g., by taking logarithms or square roots of the data) to achieve a symmetric distribution of observations that more closely resembles a normal distribution. Analyzing the means on this transformed scale would be equivalent to analyzing the medians on the original scale. We can also transform the data to their ranks, as is done for Wilcoxon's test (Chapter 7). This is the basis of a procedure known as the Kruskal-Wallis test. Finally, the analysis of variance F-test is also robust to violation of the assumption of homogeneity of variances, provided the number of replicate observations is the same for each treatment. When the various treatment samples are unequal in size, however, large differences among the variances can have a marked effect on the F-test. A preliminary test can be performed to check the assumption of homoscedasticity. If significant heteroscedasticity is present, it may be possible to transform the data to achieve homogeneity among the variances.

SUMMARY

1. The analysis of variance is a procedure for partitioning the sum of squared deviations from the mean into components associated with recognized sources of variation. The sums of squares are divided by their respective numbers of d.f. to obtain mean squares, and ratios of mean squares are compared with the F distribution to test null hypotheses that sets of means are all equal, or that certain sources of variation are not significant.

2. Every analysis of variance is based on a particular linear model, and so the analysis of variance is basically the same procedure as regression

analysis. Whereas in regression analysis the independent variables are quantitative and usually continuous, in the analysis of variance they are always discrete.

3. The appropriate analysis and F-tests depend on the experimental design. Simplest is a one-way, or among-and-within-groups, analysis of variance. The null hypothesis is that the group means are all equal, and the F-statistic is the among-groups mean square divided by the within-groups, or residual, mean square.

4. When there are multiple classifications of the data, the appropriate divisor in the F-ratio may depend on the sampling fraction—the fraction of the total number of treatments of a particular type that are represented in the study. If the sampling fraction is 1, the treatment effects are fixed; if it is less than 1, they are random. For an F-statistic to follow the F distribution under H_0, it is necessary for both the numerator and denominator to have the same expected value under H_0.

5. In a nested, or hierarchical, analysis of variance, the residual mean square is the appropriate divisor to test for differences among groups at the lowest level within groups at higher levels. The appropriate mean square to test for differences among groups at higher levels depends on the sampling fraction(s) at lower levels.

6. In the analysis of cross-classified data, such as arise in a factorial arrangement of treatments, we define simple effects (cell means), main effects (averages of cell means that pertain to one level of a factor), and interaction effects (differences of simple effects that detect "non-parallelism"). If there are no interaction effects between the factors A and B, then the effects of the different levels of A and B are additive.

7. In a two-way analysis of variance, the residual mean square is appropriate to test for interaction effects. Either the residual or the interaction mean square is appropriate to test for one of the main effects, depending on whether the other main effects are fixed (sampling fraction = 1) or a random sample from a large population (sampling fraction near 0). Use of the interaction mean square will always result in a test that is either valid or conservative, although perhaps not powerful. Sometimes interaction can be removed by a transformation of the data.

8. The analysis of covariance is a combination of regression analysis and the analysis of variance. It is often used to allow for concomitant variables when comparing several groups.

9. Analysis of variance F-tests assumes: (1) the numerator and denominator are independent; (2) we have random samples; (3) we have normally distributed populations; and (4) the variances of all the populations are equal. Assumptions (1) and (2) are satisfied by appropriate randomization. The test is fairly robust against nonnormality and also, provided the individual treatment samples are the same size, against heteroscedasticity.

FURTHER READING

Dunn, O.J. and Clark, V.A. (1974) Applied Statistics: Analysis of Variance and Regression. John Wiley & Sons (Chapters 5 to 7 give further computational details of the analysis of variance for designs we have discussed. The mathematical level of this text is fairly elementary.)

Neter, J., Wasserman, W., and Kutner, M.H. (1990) Applied Linear Statistical Models. Irwin. (This book provides a good coverage of linear models, including both regression analysis and analysis of variance.)

PROBLEMS

1. Which of the following is *not* an assumption required in using the F distribution to test hypotheses in an analysis of variance?
   A. The numerator and denominator of the F-ratio are independent.
   B. The observations represent random samples from the populations being compared.
   C. The underlying linear model is a random model.
   D. The observations are drawn from normally distributed populations.
   E. The variances of all the populations are equal.

2. If the data in an analysis of variance fail to satisfy the required assumptions, then the F-tables may suggest that the p-value is 0.05 when in fact it is
   A. $(0.05)^2$
   B. exactly 0.10
   C. less than 0.05
   D. greater than 0.05
   E. either less than or greater than 0.05

3. An investigator randomly assigned 8 patients to each of 3 different diets to study their effects on body weight. The resulting data were subjected to an analysis of variance. The F-test for the hypothesis that the mean response was the same for the 3 diet groups has degrees of freedom as follows:
   A. numerator d.f. = 2, denominator d.f. = 8
   B. numerator d.f. = 3, denominator d.f. = 7
   C. numerator d.f. = 2, denominator d.f. = 21
   D. numerator d.f. = 8, denominator d.f. = 24
   E. numerator d.f. = 7, denominator d.f. = 21

4. Consider the six treatment groups A_1B_1, A_1B_2, A_1B_3, A_2B_1, A_2B_2, and A_2B_3. Suppose the mean responses to these treatment combinations are as follows:

	B_1	B_2	B_3
A_1	10	30	40
A_2	12	33	41

   The difference 33 − 30 = 3 is
   A. the difference between simple effects of A at B_2
   B. the difference between main effects of A at B_2
   C. an interaction of A at B_2
   D. the simple effect of B_2
   E. the main effect of B_2

5. Consider the six treatment groups A_1B_1, A_1B_2, A_1B_3, A_2B_1, A_2B_2, and A_2B_3. Suppose the mean responses to these treatment combinations are as follows:

	B_1	B_2	B_3
A_1	21	31	34
A_2	33	40	41

   The average (33 + 40 + 41)/3 = 38 is called the
   A. main effect of A_1
   B. main effect of A_2
   C. simple effect of A_2 at B_1, B_2, and B_3
   D. interaction of A and B
   E. error of A

6. An investigator studies the effect of three treatments denoted A_1, A_2, and A_3 on blood pressure in patients with hypertension. These three treatments are

the only ones of interest, so inferences will pertain only to A_1, A_2, and A_3. We say the effects of these treatments are

A. fixed
B. random
C. mixed
D. additive
E. iterative

7. A mixed model is one that has both

A. fixed effects and random effects
B. simple effects and main effects
C. interaction effects and main effects
D. interaction effects and fixed effects
E. interaction effects and random effects

8. A health-management corporation has a series of 12 clinics with five staff physicians at each clinic. The corporation wishes to evaluate the differences among their clinics and physicians in managing blood pressure in hypertensive patients. Ten hypertensive patients were randomly assigned to each of two randomly selected physicians within each of four randomly selected clinics. An outline of the sources of variability, degrees of freedom, and expected mean squares in the analysis of variance resulting from the study is:

Source of Variability	d.f.	Expected Mean Square
Clinics	3	$\sigma_R^2 + 6\sigma_{P\mid C}^2 + 20\sigma_C^2$
Physicians within clinics	4	$\sigma_R^2 + 10\sigma_{P\mid C}^2$
Patients within physicians within clinics	72	σ_R^2
Total	79	

The appropriate denominator for the F-test of the hypothesis that the clinic means are all the same is

A. the calculated mean square for clinics
B. the calculated mean square for physicians within clinics
C. the calculated mean square for patients within physicians within clinics
D. the calculated mean square for the total sample
E. none of the above

9. An experiment was conducted in which patients with chronic hypertension were administered one of two doses of drug A (A_1 or A_2), in combination with one of three doses of drug B (B_1, B_2 or B_3). In all there were six treatment

groups. Forty-eight patients were randomly assigned to these groups, eight patients to a group. An analysis of variance of diastolic blood pressure after three weeks on treatment was carried out. The following table outlines the basis for an appropriate ANOVA:

Source of Variability	**d.f.**	**Expected Mean Square**
Drug A	1	$\sigma_R^2 + 24\sigma_A^2$
Drug B	2	$\sigma_R^2 + 16\sigma_B^2$
Interaction	2	$\sigma_R^2 + 8\sigma_{AB}^2$
Residual	42	σ_R^2
Total	47	

The appropriate denominator for the F-test of the hypothesis that the main effects of drug A are equal is

A. the calculated mean square for drug A
B. the calculated mean square for drug B
C. the calculated interaction mean square
D. the calculated residual mean square
E. none of the above

10. Consider the four treatment groups A_1B_1, A_1B_2, A_2B_1, and A_2B_2. Suppose the mean responses to these treatment combinations are as follows:

	B_1	**B_2**
A_1	50	52
A_2	60	73

An estimate of the interaction effect between the factors A and B is

A. 2
B. 11
C. 13
D. 22
E. 26

11. Suppose treatment A has dose levels A_1, A_2, and A_3, and treatment B has dose levels B_1 and B_2, in an experiment with a completely randomized design and a factorial arrangement of treatments. Further suppose that the differences among the simple effects of A_1, A_2, and A_3 are not the same for the two levels of B. This phenomenon is an example of

A. fixed effects
B. random effects

C. mixed effects
D. main effects
E. interaction effects

12. An analysis of variance was carried out after transforming the data to a logarithmic scale. One purpose of the transformation might have been to
A. remove single effects
B. remove main effects
C. remove fixed effects
D. remove random effects
E. remove interaction effects

13. In a study to investigate the effectiveness of an anticoagulant, 10 rats are randomly assigned to each of two groups. The rats in the first group receive an injection of the anticoagulant, while those in the second group receive a control saline injection. Samples of blood are taken from each rat before and after treatment and the coagulation time in minutes noted. The following method might be appropriately used to test the hypothesis that the mean coagulation time is not affected by treatment:
A. paired t-test
B. one-way analysis of variance of the pretreatment values
C. the correlation between pretreatment and posttreatment values
D. analysis of covariance
E. contingency table analysis

14. In performing an analysis of covariance we assume
A. interaction effects are present
B. the paired t-test is appropriate
C. parallel lines on regressing the dependent variable on the covariate
D. categorical data provide efficient estimators of the slopes for the regression lines
E. the covariate is a nominal variable

15. Analysis of covariance is often used in the statistical interpretation of experimental data to
A. increase power
B. decrease interaction
C. eliminate mixed effects
D. decrease main effects
E. increase the slope of the regression line

CHAPTER ELEVEN

Some Specialized Techniques

CHAPTER ELEVEN

Key Concepts

univariate analysis
multivariate analysis (MANOVA)
multivariate general linear models (MGLM)
discriminant analysis, discriminant function
logistic, or logit transformation, logistic regression
survival time, singly and progressively censored data, survivorship function or curve, probability density function, hazard function, proportional hazards, Cox's regression model, life-table method, Kaplan-Meier method

Symbols and Abbreviations

$S(t)$	survivorship function (cumulative survival rate)

Some Specialized Techniques

We have presented in the previous chapters basic concepts that should serve as building blocks for further study. These concepts have been illustrated by describing some of the common statistical methods found in the medical literature. It would, however, be impossible to cover in a single book all the statistical techniques that are referred to in medical journals. In this chapter we briefly familiarize you with some of the more specialized techniques of statistical analysis. Although the choice of which techniques to include and which to exclude is somewhat arbitrary, our aim has been to cover a few of the advanced methods of analysis that are more frequently encountered in medical research articles.

MULTIVARIATE ANALYSIS

Interest often centers on the simultaneous analysis of several dependent variables rather than a single dependent variable. Let us suppose, for example, that an experiment is designed to study the effect of a treatment on the following variables: diastolic blood pressure, serum cholesterol, and body weight. In particular, let us suppose that the purpose of the study is to compare the means of these three dependent variables in a treated group to the corresponding means of these three variables in a control group. If one focused on a single variable (e.g., diastolic blood pressure), then one of the methods described earlier could be used to analyze the data. All the methods we have described so far are **univariate** methods, in that only one dependent variable (variate) is involved. (Sometimes multiple regression, in which there is more than one independent variable, is also called a multivariate method. This terminology, however, is incorrect if there is only one dependent variable.) If, on the other hand, all three variables are analyzed simultaneously as dependent variables, then the analysis is termed **multivariate.** For each of the univariate methods described earlier, there is an analogous multivariate method. For example, the multivariate analogue of Student's t-test for comparing the means of two groups is called Hotelling's T^2-test, named after the American statistician Harold Hotelling (1895–1973). Similarly we can have multivariate regression analysis, multivariate analysis of variance (**MANOVA**), and multivariate analysis of covariance. Earlier, we discussed longitudinal and repeated measures data. These are special types of multivariate data and we must consider this multivariate aspect of the data in the statistical analysis. A unified approach to multivariate analysis for comparing group means is provided by **multivariate general linear models** (MGLM).

You may wonder what the advantage is to performing a multivariate analysis rather than performing a set of univariate analyses. Why not, in our example, simply perform three t-tests: one for diastolic blood pressure, one for serum cholesterol, and one for body weight? Multivariate analysis has two advantages. First, it helps overcome the problem that, as the number of statistical tests performed at a given significance level increases, so does the probability (under H_0) of finding at least one significant result. Suppose there are 20 dependent variables and we perform 20 t-tests at the 5% significance level (i.e., we decide beforehand to declare a significant finding if any one of the 20 p-values is less than 0.05). By chance alone, if there are no pairwise differences between the means, we should expect one of the tests to yield a significant result. If, however, we first perform a multivariate test, which takes account of the fact that 20 comparisons are being made, then we can appropriately control the overall probability of a type I error to be, for example, 0.05. When we perform 20 t-tests each at the 0.05 significance level, the overall probability of a type I error (i.e., the probability of rejecting the null hypothesis that all 20 pairs of means are equal when in fact they are equal) is much larger than 0.05.

A second advantage of multivariate tests is that they are more sensitive to detecting group differences that depend on certain relationships among the dependent variables. This can best be seen by considering a simple situation in which there are just two groups and two dependent variables, y_1 and y_2, illustrated graphically in Figure 11–1. Here we have graphed, for two samples of 10 study units each, the values y_1 and y_2 observed on each study unit. It is clear from this scatter diagram that the two groups are completely separate, and a multivariate test of these data would show a highly significant difference between the two groups. But if we were to perform a t-test on the 20 values of y_1, or on the 20 values of y_2, neither result would

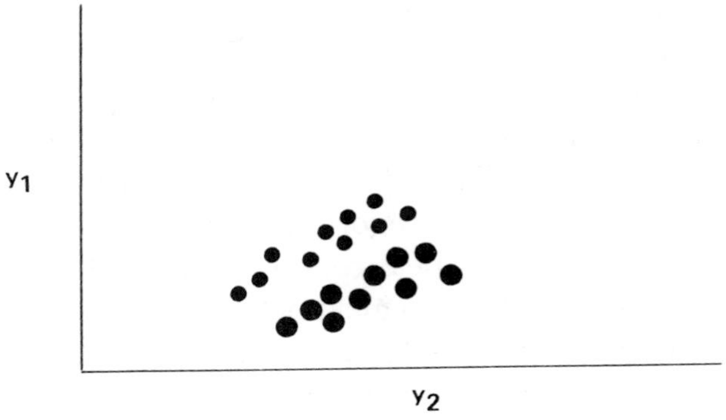

Figure 11–1. Scatterplot of the variables y_1 and y_2 measured on the ten study units in each of two groups, indicated • and ⬤.

be very significant because there is almost complete overlap between the two groups on each variable singly.

DISCRIMINANT ANALYSIS

Let us consider the situation in which it is unknown to which of two groups or populations a person belongs. For example, we may wish to know whether or not a woman is a carrier of the sex-linked hemophilia gene and hence has a risk of bearing a son with hemophilia. Suppose we have laboratory data available on the woman and we wish to use the information to classify her. Using a procedure known as **discriminant analysis,** it is possible to determine a mathematical function for this purpose, from data on a set of previously classified women. Thus we would obtain two samples of women, one of women known to carry the gene (so-called obligate carriers) and one of women known not to carry the gene. (If a woman has two hemophiliac sons, for example, she is an obligate carrier; if she has no relatives with hemophilia, on the other hand, we can be virtually certain she does not carry the hemophilia gene.) A blood sample is taken from each woman and a set of relevant measurements, such as clotting-factor levels, are determined. The result of a discriminant analysis applied to these data is a **discriminant function** that can be used to help classify a woman whose maternal uncle (but no other relative), for example, has hemophilia. In the case of hemophilia A, if we let

$$y_1 = \log \text{(clotting factor XIII coagulant activity level)}$$

and

$$y_2 = \log \text{(clotting factor XIII–related antigen level)}$$

the discriminant function derived from these two variables is (approximately) $3y_1 - 2y_2$. If this function is applied to the 20 points plotted in Figure 11–1, for example, the two groups are found to be distinct, with no overlapping of their ranges.

Discriminant analysis can also be used to classify individuals into one of several disease categories, based on vital signs, laboratory data, or both. There will usually be errors associated with such classifications, and we try to develop discriminant functions that will correctly classify individuals with a high probability. Because we cannot know whether a particular individual is classified correctly, we often estimate the probability of an individual belonging to each population. The higher the probability associated with a person belonging to a particular disease category, the more confidence we have that we can correctly classify that person.

LOGISTIC REGRESSION

In fitting a statistical model to a set of data, sometimes the dependent variable is dichotomous, whereas the independent variables are continuous, discrete, or both. In the simplest situation, we would have one dependent and one independent variable. For example, the dependent variable may be success or failure after a treatment, and we wish to model this response (the proportion of successes or failures) at selected doses of some treatment. In slightly more complex situations, we may want to model the proportion of failures (e.g., the proportion of a population with disease) in terms of suspected risk factors such as age, weight, blood pressure, etc. In cases such as these, the cumulative distributions tend to be S-shaped, or tilted S-shaped, as in Figure 11–2. This characteristic shape arises because failures often occur infrequently at low levels of the independent vari-

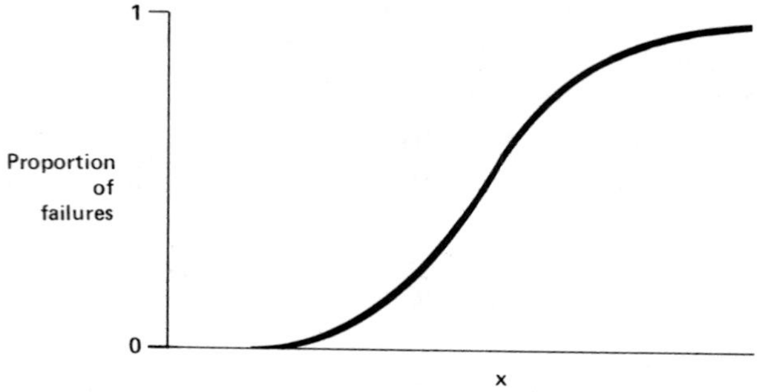

Figure 11–2. Example of a curve depicting the cumulative proportion of failures as a function of an independent variable, x.

able(s), then there is a range in which the failure rate increases rapidly, and finally there is a range in which most of the failures have occurred and so additional failures occur less frequently again.

A family of mathematical equations that has a shape resembling that of Figure 11–2 is given by the equation

$$y = \frac{1}{1 + e^{-(\beta_0 + \beta_1 x)}}$$

or, equivalently,

$$y = \frac{e^{\beta_0 + \beta_1 x}}{1 + e^{\beta_0 + \beta_1 x}}$$

With further algebraic manipulation, this is the same as

$$\frac{y}{1 - y} = e^{\beta_0 + \beta_1 x}$$

or, taking natural logarithms of both sides of this equation,

$$\log_e \left[\frac{y}{1 - y} \right] = \beta_0 + \beta_1 x$$

Thus, we have a transformation that converts the curve in Figure 11–2 into a straight line. This transformation is called the **logistic** or **logit transformation;** i.e., the logistic transformation, or logit, of y is $\log_e [y/(1 - y)]$. This is the basis of a **logistic regression** model, in which the logit of the dependent random variable Y (a proportion) is regressed on the independent variable x; i.e., the model is

$$\log_e \left[\frac{Y}{1 - Y} \right] = \beta_0 + \beta_1 x + \epsilon$$

where ϵ is a random error. There may also be several independent variables, in which case the model is of the form:

$$\log_e \left[\frac{Y}{1 - Y} \right] = \beta_0 + \beta_1 x_1 + \beta_2 x_2 + \cdots + \beta_p x_p + \epsilon$$

A variety of computer programs are available for obtaining maximum likelihood estimates of the parameters ($\beta_0, \beta_1, \ldots, \beta_p$) of this model, and for testing hypotheses about them using the likelihood ratio criterion. The use of logistic regression models is fairly common in the medical literature, especially with Y representing the probability of disease, so that $Y/(1 - Y)$ are the odds in favor of disease. Thus the $\log_e$ of the odds is estimated by the regression function $b_0 + b_1x_1 + b_2x_2 + \cdots + b_px_p$, and the odds are estimated by $e^{b_0+b_1x_1+b_2x_2+\cdots+b_px_p}$. Now suppose, for example, that $x_1 = 1$ if there is exposure to some environmental factor, and $x_1 = 0$ if there is no such exposure. Then the odds ratio for exposed versus unexposed is estimated as

$$\frac{e^{b_0 + b_1 + b_2x_2+\cdots+b_px_p}}{e^{b_0+b_2x_2+\cdots+ b_px_p}} = e^{b_1}$$

Thus b_1 is the $\log_e$ of the odds ratio for exposed versus unexposed in this example. Recall that odds ratios are particularly useful statistics for summarizing the results of case-control studies (Chapter 3). We cannot estimate the probability of disease in such studies; but, by letting Y be the proportion of cases in the study, logistic regression can be used to find the odds ratios for several different kinds of exposures (x_1, x_2, etc.). The corresponding estimated regression function (i.e., $b_0 + b_1x_1 + b_2x_2 + \ldots + b_px_p$) can also be used as a discriminant function to help classify future persons into one of the two classes (which in this instance are "disease" and "no disease").

ANALYSIS OF SURVIVAL TIMES

In some studies, especially clinical trials, the dependent variable of interest may be the amount of time from some initial observation until the occurrence of an event, such as recurrence of disease, death, or some other type of failure. This time from initial observation until failure is called the **survival time.** Statistical analysis of a group of survival times usually focuses on the probability of surviving a given length of time, or on the mean or median survival time.

A distinguishing feature of survival data is the fact that the distribution of survival times is often skewed and far from normal. Furthermore, the exact survival times of some of the study units may be unknown. For example, a group of patients may all enter a study at the same time, but some may not have "failed" by the end of the study, or they may be lost to follow-up at some point in the study. In such cases the survival times are said to be **censored.** If a study is conducted so that subjects are observed until a prespecified proportion (e.g., 60%) have failed, or if all subjects are observed for a fixed period (e.g., 5 years), the resulting survival data are said to be **singly censored.** In most clinical studies, however, patients are recruited into the study over time, and each patient is observed for a different length of time. Then, if some of the patients have not failed by the end of the study, the resulting survival data are said to be **progressively censored.**

A distribution of survival times can be characterized by one of three functions: (1) the survivorship function, (2) the probability density function, and (3) the hazard function.

The **survivorship function** S(t) is defined as the probability that a study unit survives longer than time t; i.e., if T is the random variable denoting survival time,

$$S(t) = P(T > t)$$

S(t) is also known as the cumulative survival rate, and the graph of S(t) is called the **survival curve** (Fig. 11–3). At any time t, S(t) gives the proportion still surviving at time t.

Recall that the cumulative distribution function of T is given by $F(t) = P(T \leq t)$. Hence it follows that

$$F(t) = 1 - S(t)$$

The corresponding density function is the **probability density function** f(t) of the survival time. Areas under this curve represent the probability of failure in intervals of time.

The **hazard function** of survival time T is the density of failure at a particular time, given that there has been survival until that time. The hazard function is also known as the *instantaneous failure rate*, the *force of mortality*, or the *conditional failure rate*. It can be thought of as the "instantaneous" probability of failure at a particular time given there has been survival until that time. Because time is a continuous variable, however, it is a density rather than a probability. It is equal to f(t)/S(t). In 1972 Sir David

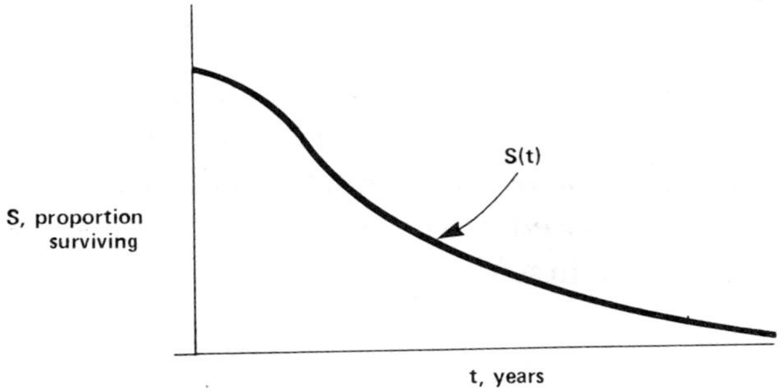

Figure 11–3. Example of a survival curve.

Cox, a British statistician, introduced a method of analyzing survival times based on the assumption that the effect of each of the factors $x_1, x_2, \ldots, x_p$ is to multiply the whole hazard function by a certain amount. The underlying model is therefore called a **proportional hazards model** or sometimes simply **Cox's regression model.** Specifically, denoting the hazard function h(t), the model can be written as

$$h(t) = h_0(t)e^{\beta_1 x_1 + \beta_2 x_2 + \cdots + \beta_p x_p}$$

where $h_0(t)$ is the hazard when all the x variables equal zero. The regression coefficients $\beta_1, \beta_2, \ldots, \beta_p$ are estimated by a maximum likelihood method that does not depend on the shape of h(t) or $h_0(t)$, and the estimates measure the effect of each factor on the hazard function. If, for example, x_2 is the amount of a particular food eaten, then the hazard function is multiplied by e^{β_2} for each unit of that food eaten; $\beta_2 > 0$ would imply that the food has a harmful effect (increasing the hazard), while $\beta_2 < 0$ would imply a beneficial effect (decreasing the hazard). Just as for logistic regression, there are computer programs for obtaining maximum likelihood estimates of the parameters and for testing hypotheses about them, in large samples, using the likelihood ratio criterion.

ESTIMATING SURVIVAL CURVES

We shall describe two methods of estimating survival curves: (1) the **life table method** and (2) the **Kaplan-Meier method.** In the life-table method, the survival times are first grouped into fixed intervals such as months or years. Let n_i be the number of study units surviving at the beginning of the ith interval, d_i the number of failures in the ith interval, and c_i the number of censored survival times in the ith interval. Then the probability that a study unit that has survived to the beginning of the ith interval will survive through to the end of that interval is estimated as

$$s_i = \frac{n_i - d_i - c_i/2}{n_i - c_i/2}$$

(It is assumed that the censored individuals leave randomly throughout the interval, so that on an average only half of them are present during the interval.) The overall probability of surviving until the end of the kth interval is estimated as the product of the probabilities of surviving through each of the first k intervals, that is

$$s_1 s_2 \ldots s_k$$

The Kaplan-Meier method of estimating survival curves uses the exact failure times rather than grouping them into intervals. Denote the ranked times of failure and censoring by $t_1 < t_2 < \ldots < t_i < \ldots < t_m$. Let u_i be the number of units surviving at time t_i, and f_i the number that fail at time t_i. A unit with survival time censored at time t_i is assumed to survive up to and including time t_i. Then the probability that a study unit that has survived to time t_{i-1} will survive to time t_i is estimated as

$$q_i = \frac{u_i}{u_i + f_i}$$

As before, the overall probability of surviving to time t_k is estimated as

$$q_1 q_2 \cdots q_k$$

Expressions are available for the standard deviations of each of these estimates of the survival times.

SUMMARY

1. Multivariate analysis is the simultaneous analysis of several dependent variables. It allows for appropriate control of the probability of a type I error when many dependent variables are involved; and it can sometimes detect group differences that are not obvious when the variables are examined individually. Every univariate method of analysis has a multivariate analogue.

2. The purpose of discriminant analysis is to find a function of several variables that can help classify a study unit as coming from a particular population.

3. Logistic regression is used to model a proportion of a population (e.g., the proportion with a disease), or a probability, as a function of one or more independent variables. The logistic transformation changes a tilted S-shaped curve into a straight line. The estimated regression coefficients can be interpreted as the logarithms of odds ratios. The estimated regression function can also be used as a discriminant function (e.g., to help classify persons as having a disease or not).

4. Survival analysis is used when the dependent variable is a survival time (i.e., the time to a well-defined event, such as death). The distribution of survival times, which is usually skewed, may be characterized by a

probability density function, a survivorship function (the complement of the cumulative distribution function), or a hazard function.

5. The hazard function gives the density of failure at a particular time, given there has been survival up to that time. In the proportional hazards model (Cox's regression model) it is assumed that the effect of each independent variable is to cause the whole hazard function to be increased or decreased multiplicatively.

6. Survival data are usually censored either singly (if every unit has been observed for the same amount of time) or progressively (if the units have been observed for different lengths of time). Two methods of estimating survival curves (that allow for censoring) are the life-table method and the Kaplan-Meier method.

FURTHER READING

Everitt, B.S. (1989) Statistical Methods for Medical Investigations. Oxford University Press. (This book gives a good overview of some of the topics in this chapter. Only a limited mathematical background is required to understand the material.)

PROBLEMS

1. An analysis is carried out to study the effects of three treatments on total serum cholesterol in patients with elevated cholesterol levels. The statistical model underlying the analysis included age of the patient as an independent variable. The resulting analysis is called
   A. analysis of covariance
   B. multivariate analysis
   C. discriminant analysis
   D. logistic regression analysis
   E. survival analysis

2. An experiment was conducted in which patients were randomly assigned to either an active treatment or a placebo. After 3 months of treatment, data were obtained for four variables: total serum cholesterol, serum triglyceride, systolic blood pressure, and diastolic blood pressure. A statistical analysis was carried out to test the null hypothesis that the treatment had no effect on any of the four variables. The resulting analysis is called
   A. univariate analysis
   B. discriminant analysis

C. logistic regression analysis
D. survival analysis
E. multivariate analysis

3. An investigator studied two groups of patients: one group with confirmed coronary heart disease and a second in which overt coronary heart disease was not present. Total serum cholesterol, serum triglyceride, systolic blood pressure, and diastolic blood pressure were determined for each patient. The investigator wished to derive from these data a mathematical function that would help decide whether a patient with unknown coronary heart disease status, but on whom these four variables had been determined, has coronary heart disease. An appropriate statistical method for doing this is
   A. univariate analysis
   B. discriminant analysis
   C. survival analysis
   D. analysis of variance
   E. analysis of covariance

4. Which of the following is an advantage of multivariate analysis?
   A. The computations for it are simpler.
   B. It requires fewer assumptions.
   C. It allows for proper control of the type I error when tests are performed on many dependent variables.
   D. It avoids the requirement of randomization.
   E. It always provides a more powerful test than a set of separate univariate analyses.

5. An investigator is studying the probability of disease in relation to several suspected risk-factor variables. A plot of the proportions with disease in various categories of each of the risk-factor variables indicates that each of the cumulative distributions is shaped like a tilted S. This suggests that the investigator should consider a
   A. univariate analysis
   B. discriminant analysis
   C. analysis of covariance
   D. logistic regression analysis
   E. none of the above

6. An investigator reported that the data from a study were analyzed using the Kaplan-Meier method. The investigator was most likely studying
   A. multivariate data
   B. survival data
   C. discrete data

D. uncensored data
E. none of the above

7. An investigator wishes to estimate the instantaneous probability that a patient will die, given that the patient has survived a given amount of time since an operation. In other words, the investigator is interested in estimating the following function of time since the operation
   A. probability density function
   B. survivorship function
   C. hazard function
   D. cumulative distribution function
   E. none of the above

8. A study of survival times of patients receiving coronary bypass operations is terminated while some of the patients are still surviving. For purposes of analysis, the survival times of these patients are said to be
   A. discrete
   B. multivariate
   C. censored
   D. terminated
   E. none of the above

9. A researcher wishes to develop a statistical model to predict serum cholesterol levels based on a knowledge of five measures of dietary intake. The method for developing such a model can be described as
   A. multiple regression analysis
   B. multivariate analysis
   C. discriminant analysis
   D. analysis by Cox's regression model
   E. survival analysis

10. An analysis is performed in which the proportion of persons with a disease in a group is divided by the proportion without the disease. A multiple regression analysis is carried out on the logarithm of the resulting ratio. This is an example of a general method known as
   A. correlation analysis
   B. multivariate analysis
   C. survival analysis
   D. logistic regression analysis
   E. censored data analysis

11. A statistician is faced with the analysis of a set of data comprising measurements of three continuous dependent variables, observed in an experiment that

used a factorial arrangement of treatments in a completely randomized design. Based on this information, the most appropriate method of analysis is

A. discriminant analysis
B. paired t-test
C. multivariate analysis of variance
D. survival analysis
E. proportional hazard function analysis

12. Cox's proportional hazards model is used to investigate relationships between survival time and a set of

A. discriminant functions
B. percentiles
C. prognostic factors
D. censored data
E. cumulative distribution functions

13. All the following are multivariate statistical techniques except

A. Hotelling's T^2-test
B. MANOVA
C. discriminant analysis
D. Student's t-test
E. MGLM

14. Subjects are recruited into a study over time as they come out of intensive care from a particular operation, and the study is terminated after 30% of the subjects have relapsed. The survival time to relapse is said to be

A. a logistic regression
B. progressively censored
C. missing
D. a maximum likelihood estimate
E. multivariate

15. A randomized, double-blind clinical trial was conducted to study the effect of a drug for lowering blood pressure versus a placebo control. The response variables of interest were systolic and diastolic blood pressure. Based on this information, the statistical analysis requires a technique appropriate for

A. data with missing endpoints
B. censored data
C. multivariate response
D. categorical response
E. noncompliance

CHAPTER TWELVE

Key Concepts

meta-analysis, effect size
multiple comparisons, Bonferroni's method
selection bias:
reporting bias
publishing bias
retrieval bias

Guides to a Critical Evaluation of Published Reports

In reading a report published in the literature, one often begins by reading the abstract or summary. While this step is important in that it quickly indicates whether the article is really of interest, its role must be kept in perspective. You must not yield to the temptation of accepting conclusions from the summary without appraising the merit and validity of the study itself. You must read the article critically before accepting its conclusions as being relevant to the way you manage patients. Experience will improve your ability to evaluate research reported in the literature, but that ability will be best utilized if you approach your reading with a definite plan in mind. A wide variety of procedures are used to conduct, analyze, and report research findings, and so it is impossible to give a single set of hard and fast rules for evaluating all such reports. We have nevertheless compiled a few guidelines that you should find helpful to keep in mind as you read the literature.

THE RESEARCH HYPOTHESIS

A first step in reviewing any article is to identify the research hypothesis. Why was this research performed? Does it have relevance to you? Is there any practical or scientific merit to it? If not, there is no need to read any further.

VARIABLES STUDIED

When you have identified the research hypothesis, and before you read the report in detail, ask yourself what variables would shed light on the research hypothesis. Next, identify the variables included in the report. Which are the dependent variables? Which are the independent variables? Are these variables relevant to the research hypothesis? Compare your list of variables with those included in the report. Were important—possibly confounding—variables overlooked? Was adequate information collected on all relevant concomitant variables? Age, race, and gender are three important variables in many human studies. Ethnic origin, geographic location, and socioeconomic status are other variables that may be important.

The methods used to obtain the data are another important consideration. Were state-of-the-art clinical or laboratory techniques used? Do these methods produce precise measurements? The scale of measurement has an impact on the choice of statistical hypotheses to be tested and statistical methods for testing these hypotheses. Hence the scale of measurement used for recording each variable should be identified; some scales may be nominal, some ordinal, and some interval.

THE STUDY DESIGN

It is imperative to identify the study design because the appropriate methods of analysis are determined by it. Is the study experimental or observational? What are the study units and how were they selected? Are they a random sample, or were they chosen merely because it was convenient to study these particular units? What was the target population? How does it compare with the study population? If, as is so often the case in experimental studies, a convenient (rather than a random) sample of study units was used, it should be described in sufficient detail for you to have some feel for how general any of the conclusions are. If the study units are patients with some disease at a particular hospital, for example, you must decide how representative these patients are. Examine the inclusion and exclusion criteria for entry into the study. Could the fact that these are patients at that particular hospital somehow make them unrepresentative? For example, does that hospital specialize in the more severe cases? Do geographic location or climatic conditions have an effect? In other words, can you reasonably use the results of this study to guide you in the treatment of your patients?

After the study units were selected, how was the study conducted? At what point was randomization used? Was an adequate control group included? Was double-blinding used to control observer bias?

SAMPLE SIZE

The greater the number of study units in an investigation, the more confident we can be of the results. Findings based on one or two subjects cannot be expected to be typical of large populations. The sample size must be large enough for the results to be reliably generalized. Statistical tests used to declare results significant or not significant depend upon the sample size. The larger the sample size, the more powerful the test and the more sensitive the test to population differences. Hence, if the sample size is enormous, even trivial population differences could be declared significant. In a small sample, on the other hand, a large difference may be nonsignificant. Especially in the case of a study that reports no significant differences, it is necessary to determine how large a difference could have reasonably been detected by the sample size used. A good study that reports negative results will also quote the power of the statistical tests used.

COMPLETENESS OF THE DATA

Clinical studies invariably suffer from the problem that some patients have missing data for one reason or another. If an unusually large number of study units, say more than 20%, have a significant amount of incomplete data, then the credibility of the results should be questioned. Be careful of reports in which a portion of the data have been discarded as outliers because of gross errors or other unusual circumstances. If an investigator discards those data that do not support the research hypothesis, the scientific objectivity is lost. A good study will include all the data available in the various analyses performed.

APPROPRIATE DESCRIPTIVE STATISTICS

An overview of the findings of a study can often be gleaned by scanning the tables and graphs. Be watchful, however, that the tables or graphs are not misleading. Distinguish between the sample standard deviation and the standard error of an estimated parameter. If it is not reported, calculate the coefficient of variation for some of the important variables. If the relative variability is large (for example, if the coefficient of variation is greater than 30%), then important population differences may be obscured by the "noise" in the data. Be sure you understand what the numbers in the tables represent and exactly what is graphed. It sometimes helps (especially in a poorly written report) to reconcile numbers in the tables and graphs with numbers in the text.

APPROPRIATE STATISTICAL METHODS FOR INFERENCES

The names of statistical tests used in the analysis, such as Student's t-test, paired t-test, analysis of variance, multiple regression analysis, and so forth, should be stated in the methods section of the report. Be wary of reports that state p-values without indicating the specific method used to analyze the data. In each case, identify the specific null hypothesis that is being tested. Try to determine if the statistical methods used are appropriate for the study design, scale of measurement, etc. If the method of analysis is unfamiliar to you, consult a statistician. Each method requires certain assumptions of the data, such as independent samples, random samples, normal distributions, or homogeneous variances. Gross violations of these assumptions may bias the analysis. The report of a careful analysis will justify the use of each statistical test used, should there be any doubt. Remember that a p-value represents the chance that a difference in the sample data is the result of random variation, when in fact there is no difference in the populations from which the samples came. A p-value of 0.05 tells us there is a 5% chance that the observed difference (or a more extreme difference) could arise by chance if the null hypothesis is true. Suppose three independent statistical tests are carried out. If in each case the null hypothesis is true, the possibility that at least one of the tests results in significance at the 5% level is about 0.14 $[= 1 - (1 - 0.05)^3]$. If the three tests are not independent, then the probability is somewhere between 0.05 and 0.14. If c comparisons are made, independent or not, the probability of obtaining, by chance, at least one significant at the α level is less than or equal to $1 - (1 - \alpha)^c$, which is approximately equal to $c\alpha$ for small values of α. In other words, if we want to correct a quoted p-value, p^*, for the fact that c statistical tests have been performed, this can be done conservatively by multiplying it by c; we can be sure that the appropriate p-value is less than cp^*. This approach to making **multiple comparisons** is known as **Bonferroni's method.** Suppose, for example, that a sample of patients with a particular disease is compared with a sample of controls with respect to a panel of 50 HLA antigens, in order to determine if any of these antigens are associated with the disease. If any one of the tests results in a p-value of 0.001 ($= 0.05 \div 50$) or less, overall a significant result has been found at the 0.05 level.

LOGIC OF THE CONCLUSIONS

Above all, remember that there is no substitute for evaluating the logic of the conclusions. A report that concludes it is safer to drive at high speeds because relatively few deaths from automobile accidents occur at speeds in excess of 100 miles per hour is clearly absurd. The frequency of deaths occurring at speeds in excess of 100 miles per hour needs to be related to the number of cars driven over 100 miles per hour, and this compared with some similar fraction for cars driven at slower speeds. Equally absurd conclusions, however, can be found in many research reports because improper—or no—comparisons are made. Always be on your guard against this type of fallacy; it is more common than you may suspect.

META-ANALYSIS

Finally, you should be aware of a special type of report, which is more and more commonly published in the health-care literature, called a **meta-analysis.** This refers to the statistical analysis of information from a series of studies carried out for the same general purpose by different investigators or, in some instances, by the same investigators at different times. The aim of a meta-analysis is to synthesize the findings from different sources into an overall interpretation to guide practitioners. Thus, a meta-analysis attempts to combine all the available information on a given topic by pooling the results from separate studies.

From a purely statistical point of view, the most obvious benefit of a meta-analysis is that it effectively increases the sample size and, therefore, the power of the analysis to detect important group differences that may go undetected in small individual studies. The increased sample size will also lead to more precise estimates, and, because each study will have been performed under slightly different conditions or in somewhat different populations, a meta-analysis can give evidence for the *generalizability* of a particular treatment. However, the results of a series of investigations of a particular topic may differ, and we shall see that this may present either an opportunity or some difficulty in the overall interpretation. The combined analysis should include a critical evaluation of the design, analysis, and interpretation of the individual studies that are summarized. The subjects in the different studies may be heterogeneous, treatments may vary in dosage and compliance, and experimental skills and techniques may differ.

A meta-analysis begins with an effort to identify and obtain information on all studies relevant to the topic being investigated. The extent to which this can be done depends on the methods we use to search for published and unpublished studies and on our ability to identify clearly the focus of each candidate study. Three types of **selection bias** may occur. **Reporting bias** occurs when investigators fail to report results—for example, because they did not lead to statistical significance. **Publishing bias** occurs when journal editors decline to publish studies, either in whole or in part, because of lack of statistical significance. **Retrieval bias** occurs when the investigator conducting the meta-analysis fails to retrieve all the studies relevant to the topic of interest.

Any meta-analysis is necessarily limited by what is available in the studies retrieved. Sometimes individual investigators may be willing to share additional aspects of data when published results are lacking in detail. Studies may differ in design, quality, outcome measure, or population studied. They may vary from blinded, randomized, controlled trials to trials that do not use blinding, randomization, or controls. Criteria for including studies in a meta-analysis should be well defined before the search for candidate studies begins.

Studies done at different times or by different investigators are often taken to be statistically independent. However, although working separately, investigators of a specific topic may have similar backgrounds and prior beliefs, communicate frequently with each other, and modify later studies on the basis of earlier outcomes. Thus a meta-analysis should investigate time, location, and investigator effects on the outcome measure.

Before the results of studies are combined, there must be some assessment of the homogeneity of the results. It is expected that the results of different studies will vary somewhat, but for investigations that are essentially similar, heterogeneity in the results must decrease our confidence in any conclusions. An appropriate meta-analysis will then require special statistical techniques to reflect this heterogeneity, by quoting a larger standard error for any estimates of how well a treatment performs. Alternatively, it may be that variation in outcomes across studies is the result of differences in treatment protocols; in this situation an opportunity may exist to determine the best treatment regimen.

Once it has been decided to combine the results of several studies to obtain a single overall result, it is necessary to choose whether to weight each study the same or differently. Some attempts have been made to weight studies by their relative quality, but this is subjective and very difficult to quantify reliably. Another approach is to conduct separate analyses for groups of studies of similar quality. Often, the best approach is to weight the studies according to sample size, with the larger studies weighted more heavily than the smaller ones. A good meta-analysis will use more than one weighting scheme, to be sure that the resulting tests and estimates are reasonably robust to which particular weighting scheme is chosen.

We have already seen a simple method of combining studies when we wish to perform an overall test of a null hypothesis; this is the method of combining p-values discussed in Chapter 8. Other methods are available and are discussed in the additional readings suggested at the end of this chapter. Analysis of variance techniques are often used to obtain pooled estimates, and these are similarly discussed in the additional reading. The important thing to remember is that if there is heterogeneity among studies, as indicated by a large interaction between treatments and the different studies, then this source of variation must be included in the standard error when computing an overall confidence interval for the treatment effect.

Finally, a concept often used in meta-analysis is that of **effect size.** This is defined for each candidate study as the estimated difference in mean outcome between those treated and the controls, divided by the estimated control standard deviation. This is often used to combine the effects of studies that measure different outcomes, such as might result from the use of different measuring instruments. Those studies using instruments that have larger measurement variability thus have less impact than those studies using more precise instruments.

SUMMARY

1. Do not accept conclusions solely on the basis of the abstract or summary of a published report.

2. Once you have determined the research hypothesis, determine whether all important relevant variables were studied, and whether they were studied appropriately.

3. Identify the study design and the study population to determine the relevance of the results.

4. Take account of the sample size and the completeness of the data, especially if the study reports that no significant differences were found.

5. Be sure you understand what the numbers in the text and tables represent, and what is graphed.

6. Try to determine if the statistical methods used are appropriate. If multiple significance tests are performed, multiply each p-value by the number of tests performed to obtain an upper bound for the overall significance level.

7. When reading a meta-analysis, note what steps were taken to reduce selection bias—reporting bias, publishing bias, and retrieval bias.

8. A meta-analysis must assess the homogeneity of the studies it includes. Criteria for inclusion should have been defined prior to beginning the search for candidate studies.

9. A meta-analysis can result in increased power, precision, and generalizability. But it should be demonstrated that the results are robust and do not depend critically on the particular weighting scheme used.

10. The effect size of a study—(treatment average − control average) ÷ control standard deviation—is used to pool the results of studies that differ in measurement variability.

11. Whenever you read a report, whether a single study or a meta-analysis, ask yourself if the conclusions make sense.

FURTHER READING

Haines, S.J. (1981) Six Statistical Suggestions for Surgeons. Neurosurgery 9:414–417. (Some basic principles are given for interpreting statistical analyses in the medical literature. Catchy subheadings make this article enjoyable to read.)

Glantz, S.A. (1980) Biostatistics: How to Detect, Correct and Prevent Errors in the Medical Literature. Circulation 61:1–7. (This article was referred to in Chapter 1; now is the time to read it.)

National Research Council Committee on Applied and Theoretical Statistics (1992) Combining Information: Statistical Issues and Opportunities for Research. Washington,

D.C.: National Academy Press. (This book contains a comprehensive overview of the methods of meta-analysis, together with examples from the health-care field.)

Schmidt, J.E., Koch, G.G., and LaVange, L.M. (1991) An Overview of Statistical Issues and Methods of Meta-Analysis. Journal of Biopharmaceutical Statistics 1(1):103–120. (This is a good introduction to the various statistical strategies used in meta-analysis.)

PROBLEMS

1. Read articles in your own area of interest critically, following the guidelines of this chapter.

2. Read and critique the design and analysis aspects of the studies in the following two brief papers:

   Gash, A., and Karliner, J.S. (1978) No Effect of Transcendental Meditation on Left Ventricular Function. Annals of Internal Medicine 88:215–216.

   Michaels, R.M., Nashel, D.J., Leonard, A., Sliwinski, A.J., and Derbes, S.J. (1982) Weekly Intravenous Methotrexate in the Treatment of Rheumatoid Arthritis. Arthritis and Rheumatism 25:339–341.

Epilogue

This book is different from many other books on biostatistics in that many detailed calculations and formulas have been excluded from the body of the text because they are not helpful in understanding the basic concepts involved. We have not stressed the details of calculating a regression line, for example, because the necessary calculations give little insight into the meaning of the estimates calculated. We have, however, included details of those calculations that will give the reader a better feel for the underlying concepts. Thus we have carefully explained how to calculate a sample standard deviation and a sample correlation coefficient, because these are cases in which going through the actual calculations will give the reader a better feel for what the estimates measure.

This book is also different from many introductory books in that it includes some topics that are usually covered only in mathematically more sophisticated texts. We have included, for example, the difference between estimates and estimators, maximum likelihood estimation, the likelihood ratio criterion, and random effects models in the analysis of variance. We have described both hypothesis testing, a procedure for making a decision between two hypotheses, as is usually done, and significance testing, a means of quantifying disbelief in the null hypothesis by determining a p-value. Some statisticians may disagree with the choice of topics we have included as "essentials" of biostatistics, and with the way in which we have covered them. Now that you have read this book, however, you should have a basic understanding of most of the more common statistical methods used, and you should be in a good position to learn about many more that you may come across. Statistical methods are becoming more and more sophisticated every day. It would be impossible to cover in a book this size all the various statistical tests that occur in the literature of biomedical research. If a report you are reading refers to a particular statistical method you do not know about, consult a more advanced book or seek the help of a biostatistician who can explain the essentials of the method in simple terms.

If you need to use some of the simpler statistical methods, you may well be able to do so yourself with the help of this and/or other statistical textbooks. One need not always be a surgeon to remove a splinter. But, just as you would entrust major surgery to a specialist, so should you consult a competent biostatistician for major statistical work. Similarly, just as you can expect the surgeon to explain the surgery that will be performed, so can you expect the biostatistician to explain the statistical analysis that will be performed. You do not need the gory details, but you should understand the essentials. Of course there is nothing to stop you from learning more and perhaps becoming one day yourself a competent biostatistician, should you so desire. The purpose of our book, however, has been merely to help you understand the statistical principles underlying much of medical practice and the statistical content of biomedical research articles. If, in addition, this has the ultimate effect of upgrading the quality of statistical usage in your area of interest, we shall have been more than repaid for our efforts.

Review Problems and Answers

REVIEW PROBLEMS

1. Statistics is used in medical research
   A. to make rational inferences
   B. to quantify uncertainty
   C. to summarize the results of experiments
   D. all of the above
   E. none of the above

2. The following are steps in the scientific method—the never-ending circle of refining our knowledge about the universe:

   (a) test a hypothesis by experiment
   (b) formulate hypothesis
   (c) retain or reject hypothesis
   (d) observe

   The order of these steps should be
   A. dcba
   B. cbda
   C. adbc
   D. abcd
   E. dbac

3. The logic of inductive reasoning is an aspect of statistics that is used
   A. to prove mathematical theorems
   B. to make rational inferences about what happens in general
   C. to deduce specific results from general laws
   D. all of the above
   E. none of the above

4. You wish to conduct a clinical trial to compare two drugs in the treatment of a disease, and decide to enlist the help of a statistician. What is the best time to initiate contact with the statistician?
   A. before you have decided which two drug regimens to use, so that the statistician can help decide on what is medically appropriate
   B. after you have decided on what drug regimens to use, but before you decide on the sample of patients to receive them, so that the statistician can help design the trial
   C. after you have decided on the sample of patients, but before you actually conduct the trial, so that the statistician can help organize the data
   D. after the data have been collected, but before analyzing the data, so that the statistician can be responsible for the data analysis
   E. after the data are analyzed, but before the results are submitted for publication, so that the statistician can verify that an appropriate statistical analysis has been performed

5. A physician develops a diagnostic test that is positive for 95% of the patients who have the disease and for 10% of the patients who do not have the disease. Of the patients tested, 20% have the disease. If the patient's test is positive, the probability a patient has the disease is approximately
   A. 0.10
   B. 0.30
   C. 0.50
   D. 0.70
   E. 0.90

6. A 99% confidence interval implies that
   A. the probability the given interval contains the true parameter is 0.99
   B. the probability that 99% of the observations are in the interval is 0.99
   C. on average, 99 out of 100 similarly constructed intervals would contain the true parameter
   D. the probability the given interval does not contain the true parameter is 0.99
   E. there is a 1% chance the hypothesis is false

7. An investigator collects diastolic blood pressure levels on a group of patients. He divides his scale of measurement into intervals of 5 mm Hg (i.e., 70–74 mm Hg, 75–79 mm Hg, 80–84 mm Hg, etc.). The investigator counts the number of patients with diastolic blood pressures in each interval. If the investigator were to plot the frequency of blood pressure levels in each interval, he would probably choose the following type of graph:
   A. histogram
   B. scatter diagram

C. regression line
D. bar graph
E. correlation coefficient

8. The figure below depicts a distribution that is
A. symmetric
B. unimodal
C. leptokurtic
D. positively skewed
E. negatively skewed

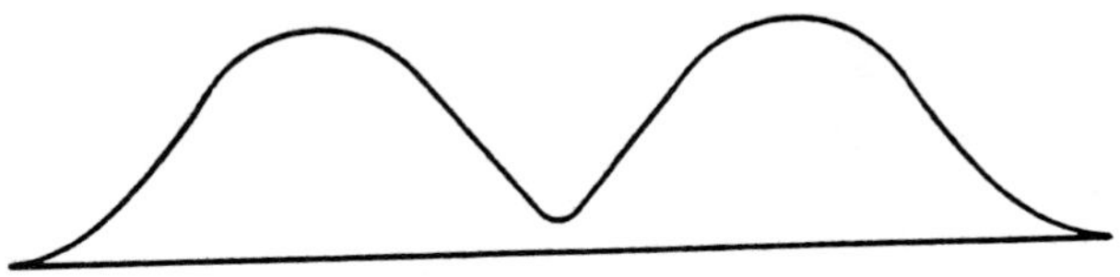

9. An investigator states that the correlation between two variables is not statistically significant (r = 0.07). Which of the following conclusions is appropriate?
A. 0.07 should be quoted as the significance level.
B. 0.07 indicates a strong linear association between the two variables.
C. For every unit of change in one variable, the other variable increases by 0.07 units.
D. High values of one variable tend to be associated with low values of the other variable.
E. Any association between the two variables does not appear to be linear.

10. Serum cholesterol levels in a group of young adults were found to be approximately normally distributed with mean 170 mg/dl and standard deviation 8 mg/dl. Which of the following intervals includes approximately 95% of serum cholesterol levels in this group?
A. 160–180 mg/dl
B. 162–178 mg/dl
C. 150–190 mg/dl
D. 154–186 mg/dl
E. 140–200 mg/dl

11. The standard deviation of a population is about 25. The standard error of the mean of a random sample of nine observations is about
A. 3
B. 8
C. 75

D. 225
E. 625

12. A difference is declared significant at the 5% significance level. This implies that
A. the difference is significant at the 1% level
B. there is a 95% probability that the difference is attributable to sampling variability
C. the difference is significant at the 10% level
D. the probability is 95% that the true difference is greater than zero
E. there is a 5% probability that there is no difference

13. An estimator that on average gives the same value as the parameter being estimated is said to be
A. minimum variance
B. maximum likelihood
C. efficient
D. unbiased
E. symmetric

14. An investigator compares two treatments, A and B, and finds that the difference in responses for these treatments is not statistically significant ($p = 0.25$). This implies that
A. the difference could well have occurred by chance alone
B. the probability the treatments are different in their effectiveness is 0.25
C. the probability the treatments are equally effective is 0.25
D. one of the treatments is 25% more effective than the other
E. the difference in success rates using the two treatments is 25%

15. A patient checks her diastolic blood pressure at home and finds her average blood pressure for a 2-week period to be 84 mm Hg. Assume her blood pressure to be normally distributed with a standard deviation $\sigma = 5$ mm Hg. A nurse checks the patient's diastolic blood pressure in the clinic and finds a value of 110 mm Hg. The clinic reading is apparently
A. not atypical of her distribution of blood pressures
B. consistent with normotensive diastolic blood pressure
C. below the 95th percentile for this patient
D. extremely high for this patient
E. none of the above

16. "p-values" are reported often in the medical literature, yet their meaning is not always understood. The p-value is
A. the power of the test
B. the probability of getting a result as extreme or more extreme than the one observed if the null hypothesis is false

C. the probability the null hypothesis is true
D. the probability of making a type II error
E. the probability of getting a result as extreme or more extreme than the one observed if the null hypothesis is true

17. An investigator found drug A to be more effective than drug B in reducing blood pressure ($p = 0.02$). A review of the literature revealed two other studies reported similar results, one finding $p = 0.03$ and the other $p = 0.01$. The probability that all three results could have occurred by chance when in fact there was no difference between the drugs is
A. 0.000006
B. 0.06
C. 0.94
D. 0.000094
E. 0.04

18. A type II error
A. is often made when the p-value is small
B. is always made when the p-value is large
C. can only be made if the null hypothesis is true
D. can only be made if the null hypothesis is false
E. none of the above

19. An investigator wishes to test the hypothesis that the variance of serum cholesterol levels in a group of school children is 20 $(\text{mg/dl})^2$. Since the serum cholesterol levels in these children appear to be normally distributed, the appropriate statistical distribution to use in evaluating the test statistic is the
A. normal distribution
B. t-distribution
C. F-distribution
D. binomial distribution
E. chi-square distribution

20. Below are five tables giving the frequency of the presence of symptoms among 200 patients with a particular disease and 200 patients without that disease. The rows and columns are: D = disease present; $\overline{D}$ = disease absent; S = symptoms present; and $\overline{S}$ = symptoms absent. Which table presents results that would be expected to yield the smallest chi-square value for testing the hypothesis that the proportion of patients with symptoms present is the same whether or not disease is present?

A.

	D	$\overline{D}$	*Total*
S	100	100	200
$\overline{S}$	100	100	200
Total	200	200	400

B.

	D	$\overline{D}$	*Total*
S	100	175	275
$\overline{S}$	100	25	125
Total	200	200	400

C.

	D	$\bar{D}$	*Total*
S	175	100	275
$\bar{S}$	25	100	125
Total	200	200	400

D.

	D	$\bar{D}$	*Total*
S	25	100	125
$\bar{S}$	175	100	275
Total	200	200	400

E.

	D	$\bar{D}$	*Total*
S	100	25	125
$\bar{S}$	100	175	275
Total	200	200	400

21. The diastolic blood pressures of a random sample of 25 men are measured. The sample mean is found to be 85 mm Hg, the standard deviation 5 mm Hg. What are the approximate 95% confidence limits for the mean diastolic blood pressure of the population sampled?
    A. 85 ± 1 mm Hg
    B. 85 ± 2 mm Hg
    C. 85 ± 5 mm Hg
    D. 85 ± 10 mm Hg
    E. 85 ± 15 mm Hg

22. Finding a p-value in significance testing begins with the basic assumption that
    A. well-trained observers record the data
    B. modern computing facilities are available
    C. the null hypothesis is true
    D. the sample size is at least 30
    E. the data are recorded on an interval scale

23. A researcher wishes to study the mean effects of four treatments, say A, B, C, and D, on triglyceride levels. He randomly assigns 40 patients to the four groups—10 patients to each group. He administers the treatments for 2 months and then measures the triglyceride level of each patient. His statistical hypothesis is that the mean triglyceride level is the same in each of the four groups. He knows from a plot of the data that, in these patients, log triglyceride levels are approximately normally distributed. The investigator should take logs of his data and use the following to test his hypothesis:
    A. F-test
    B. χ^2-test
    C. two-sample t-test
    D. paired t-test
    E. none of the above

24. Which of the following is a statistic?
    A. the population standard deviation of height
    B. the mean of a binomial distribution

C. the mean of a normal distribution
D. the population variance of height
E. the mean height of a sample of 10 men

25. The power of a test is the probability of
A. accepting a true hypothesis
B. rejecting a true hypothesis
C. accepting a false hypothesis
D. rejecting a false hypothesis

26. All of the following are based on statistical data except
A. smoking increases the risk of lung cancer
B. wearing seat belts increases the chance of survival in automobile accidents
C. the chance a newborn baby is female is slightly less than 50%
D. the sun will rise tomorrow
E. π is the ratio of the circumference of a circle to its diameter

27. Sometimes a one-tail test is used to determine significance level, and sometimes a two-tail test is used. The two-tail test is used for situations in which we know *a priori*
A. two samples are involved
B. both treatments will either increase or decrease the response
C. a paired t-test is appropriate
D. any true difference could be in either direction
E. two proportions are unequal

28. Assume the risk of developing stomach cancer during one's life is 1 in 50, whereas that of developing skin cancer is 1 in 25. Further assume that these events are independent. The risk of developing stomach cancer for a man who already has skin cancer is
A. 1 in 25
B. 1 in 50
C. 3 in 50
D. 1 in 1250
E. about 1 in 1.5 million

29. A distribution has a long tail to the right, so that it is not symmetric. We say the distribution is
A. abnormal
B. positively skewed
C. negatively skewed
D. bell shaped
E. bimodal

30. The probability an operation is a success is 3 in 4. The operation is performed on each of three patients. The probability that at least one of the operations is a failure is
    A. 1/64
    B. 63/64
    C. 27/64
    D. 37/64
    E. 48/64

31. In a study of stomach cancer, patients who had the disease were matched with patients without cancer (controls) by age, sex, and social class. The frequency of alcohol consumption was then compared. What type of study was this?
    A. clinical trial
    B. prospective
    C. experimental
    D. sample survey
    E. retrospective

32. A physician wishes to evaluate the effectiveness of three treatments, A, B, and C, with regard to time until relief of a headache. He decides to study 60 patients with chronic headache problems by giving 20 patients A, 20 patients B, and 20 patients C, and observing them until they are relieved of their headaches. The physician randomly assigns A, B, or C to the 60 patients, using the restriction that 20 patients receive each treatment, but no other restriction. The design of this study is
    A. double-blind
    B. changeover
    C. completely randomized
    D. latin square
    E. randomized blocks

33. Advantages of using the double-blind procedure in clinical trials include
    A. to increase the power of the trial
    B. to reduce observer variability in recording data
    C. to prevent biased observation of the outcome of a treatment
    D. to eliminate chances of a disproportionate number of poor-risk patients in one of the groups
    E. to allow for multiple comparisons

34. For a sample of n observations, we calculate the sample mean and then the deviation of each observation from the mean. These deviations are summed, and the result is divided by $n - 1$; the result is
    A. zero
    B. the sample mean

C. the standard error of the mean
D. the sample standard deviation
E. the sample variance

35. Three new cases of a certain disease occurred during the month of July. If 500 persons were at risk during July, then the
A. prevalence was 3 per 1,000 persons
B. incidence was 3 per 1,000 persons
C. prevalence was 6 per 1,000 persons
D. incidence was 6 per 1,000 persons
E. odds ratio was 3:1

36. A group of patients was examined during a routine screening for elevated blood pressure. Twenty patients were told they had high blood pressure. A drug was prescribed for these patients, and they were asked to return for re-examination 1 week later. At the second examination it was determined that the mean blood pressure for these patients was 10 mm Hg lower than on the initial screening. It was claimed that the drug was responsible for the decrease. However, at least part, if not all, of the decrease would be expected to be due to a phenomenon known as
A. observer bias
B. double blinding
C. random allocation
D. false-negative testing
E. regression toward the mean

37. A physician conducted a sample survey of residents who graduated from a medical school during the last 5 years. The sample design was structured so that a random 10% of the graduates from each of the five classes were interviewed. The design can be described as a
A. simple random sample
B. stratified random sample
C. systematic random sample
D. two-stage random sample
E. random cluster sample

38. A control patient of the same age, race, and sex was found for each member of a group of 25 test patients. The control patients were given a standard drug and the test patients were given a new drug to determine whether the new drug increased the number of hours slept during one night of observation. An appropriate test for the hypothesis of interest is the
A. binomial test
B. paired t-test

C. two-sample t-test
D. F-test
E. chi-square test

39. An investigator studies the effect of a drug for lowering serum cholesterol levels. Patients are randomly assigned to either an active treatment group or a placebo group. The patients' cholesterol levels are then observed for 1 year while they take one of the treatments daily. This investigation can be described as a
A. historical prospective study
B. case-control study
C. clinical trial
D. retrospective study
E. robust study

40. The incidence of a certain disease among smokers was found to be 20 per 100,000 per year, while among nonsmokers it was found to be 5 per 100,000 per year. This implies that
A. the risk of developing the disease is four times greater among smokers than among nonsmokers
B. smokers require four times as many hospital beds as nonsmokers with the disease
C. the prevalence of smoking was four times as great among diseased persons as compared to nondiseased persons
D. the risk of developing the disease appears to be unrelated to smoking
E. a clinical trial is needed to estimate the risk of developing the disease

41. In experimental studies, known sources of extraneous variability in the dependent variable are best controlled by using
A. randomization
B. completely randomized designs
C. randomized block designs
D. sample surveys
E. double-blind procedures

42. Investigator A claims his results are statistically significant at the 5% level. Investigator B argues that significance should be announced only if the results are statistically significant at the 1 % level. From this we can conclude
A. it will be more difficult for Investigator A to reject statistical null hypotheses if he always works at the 5% level (compared with Investigator B)
B. it will be less difficult for Investigator A to reject statistical null hypotheses if he always works at the 5% level (compared with Investigator B)

C. if Investigator A has significant results at the 5% level, they will also be significant at the 1% level
D. if Investigator A has significant results at the 5% level, they will never be significant at the 1% level
E. none of the above

43. The duration of disease A is longer than that of disease B. They both have the same incidence. The prevalence of disease A would then be expected to be
A. the same as that of disease B
B. less than that of disease B
C. greater than that of disease B
D. less than the incidence of the disease
E. greater than the incidence of the disease

44. Two laboratory methods for determining triglyceride levels are being compared. Method A has a coefficient of variation of 8%, while method B has a coefficient of variation of 3%. This implies that
A. method B is less precise than method A
B. the distribution of method B is less skewed than that of method A
C. method B is less apt to produce false positives
D. method B is easier to carry out than method A
E. there is less bias in method A than in method B

45. Suppose the regression equation for predicting y from x is given by $\hat{y} = 10 + 3x$. Then all the following are true except
A. the intercept is 10
B. the correlation between x and y is negative
C. the slope of the regression line is positive
D. the predicted value of y for $x = 4$ is 22
E. y increases as x increases

46. An estimator that has optimal properties when certain assumptions are true but also performs well when the assumptions are false is said to be
A. efficient
B. unbiased
C. unique
D. maximum likelihood
E. robust

47. A researcher carries out an analysis of variance on a set of data and finds a statistically significant interaction between two factors of interest. A similar analysis, performed after taking logarithms of the observed data and carried out on these transformed data, did not result in a statistically significant interaction.

The analysis of the transformed data is the preferred analysis because, on the transformed scale,

A. the effects of the two factors are additive
B. the residual errors are uniformly distributed
C. the residual errors are not independent
D. the residual errors have variances proportional to the magnitude of the difference between the means for the individual factors
E. none of the above

48. An analysis of variance was carried out to compare the mean diastolic blood pressure in four groups of patients. This resulted in the finding that the variability among groups was not significantly greater than that within groups. A closer study of the data revealed, however, that the age distribution was not the same in the four groups. An analysis of covariance was therefore performed, with age as the covariate, and then statistically significant differences were found among the groups. The most likely explanation for this disparity in the results of the two analyses is

A. the residual errors are not independent
B. nonparametric tests are more powerful than their parametric counterparts
C. more precise comparisons among the groups are possible when the obscuring effect of age is accounted for
D. p-values are easier to obtain using the analysis of covariance procedure
E. none of the above

49. The main purpose of discriminant analysis is

A. to adjust for concomitant variables
B. to remove the obscuring effects of interaction terms
C. to explain linear relationships with covariates
D. to compare categorical data
E. to classify individuals into categories

50. The estimated probabilities of survival for the first, second, and third years after surgery in a group of experimental patients were found to be 0.80, 0.67, and 0.52, respectively, while those for a control group were found to be 0.76, 0.59, and 0.37. The difference between the two survival profiles was found to be statistically significant. Based on these data, the most appropriate conclusion is

A. the experimental group appears to have a better survival profile
B. the control group appears to have a better survival profile
C. the survival profiles are about the same
D. a regression analysis is needed to compare the two groups
E. incidence rates are required to compare the two groups

ANSWERS

(The numbers in parentheses indicated for each problem refer to the relevant pages in the text.)

END-OF-CHAPTER PROBLEMS

Chapter 2

1. C (17)
2. D (17)
3. C (18)
4. D (20)
5. B (20–21)
6. C (20–21)
7. E (22–23)
8. C (22–23)
9. C (23–24)
10. E (23–24)
11. B (25–26)
12. D (25–26)
13. C (26)
14. B (25)
15. B (12–13, 27)

Chapter 3

1. B (36)
2. D (48–49)

3. (1) B (46–47)
3. (2) C (46–47)
4. C (46–47)
5. A (48)
6. A (49–50)
7. E (49–50)
8. B (51)
9. B (51, 54)
10. C (52–55)
11. C (53–54)
12. C (53–54)
13. B (53–54) Note that the range for set 1 is 30, while for set 2 it is 70.
14. A (51–52) The distribution is positively skewed; both median and mean are greater than the mode.
15. B (42–43, 51–52) The cumulative plot corresponds to the data shown in the histogram in Problem 14.

Chapter 4

1. D (68–70) $0.007 + 0.020 - 0.0008$
2. A (70, 73) $0.0008 \div 0.020$
3. E (70, 73) $0.0008 \div 0.007$
4. B (71)
5. D (71) $0.3 \times 0.3 \times 0.3$
6. B (71) We require the probability that A does not occur in the first three runs and does occur in the fourth run, i.e., $(1 - 0.3)^3 \times 0.3$.
7. E (71) $1 - 0.2 \times 0.1 \times 0.3$
8. E (45, 68–71) $\frac{1}{3} + \frac{2}{3} \times \frac{2}{3}$ or $1 - \frac{2}{3} \times \frac{1}{3}$
9. C (45, 68–71) 0.7×0.8 or $1 - (0.3 + 0.7 \times 0.2)$
10. D (75–77) P(colorblind and male) = P(male)P(colorblind | male) = $0.5 \times 0.05 = 0.025$; P(colorblind and female) = P(female)P(colorblind | female) = $0.5 \times 0.0025 = 0.00125$; therefore P(male | colorbind) = $0.025 \div (0.025 + 0.00125)$.
11. B (84) Child receives O from father; the likelihood ratio is thus $1 \div 0.67$.
12. D (68–71) P(next child will have Down) = P(translocation)P(next child will have Down | translocation) + P(no translocation)P(next child will have Down | no translocation) = $0.05 \times 0.16 + 0.95 \times 0.01 = 0.0175$.

13. C (75–77)
14. E (84)
15. A (84)

Chapter 5

1. C (92)
2. E (91–92)
3. B (95)
4. A (96) $(1 - 0.02)^4$
5. C (96) $1 - (1 - 0.25)^3$
6. B (97–99) The number of possible mutant colonies on a plate is indefinitely large.
7. B (99–100)
8. A (101–102)
9. D (101–102)
10. B (102) $15 \pm 2 \times 3$
11. B (102) 150 ± 10
12. B (102)
13. D (105–106)
14. D (105–106)
15. D (103–104)

Chapter 6

1. D (115–116) Note that the word ''estimate'' is commonly used in place of the more appropriate word ''estimator.''
2. D (114–115) $3 = s_{\bar{Y}} = s_Y \div \sqrt{n} = s_Y \div \sqrt{9} = S_Y \div 3$; therefore, $9 = s_Y$.
3. D (117)
4. D (118)
5. D (117)
6. C (123–124)
7. A (125–126) Compare t-values in the columns headed 99.5% and 97.5% in Table 2, Appendix 2.
8. C (119–120, 123–124)
9. B (123–124)
10. A (125–126)

11. A (125–126) $120 \pm 2(10 \div \sqrt{100})$
12. B (123–124) For a large ($n > 30$) sample, the width of the 95% confidence interval is approximately $4s_{\bar{Y}}$, that is, $\bar{y} + 2s_{\bar{Y}} - (\bar{y} - 2s_{\bar{Y}})$, and so is proportional to $s_{\bar{Y}}$. To halve the interval we need to halve $s_{\bar{Y}}$. Since $s_{\bar{Y}} = s_Y \div \sqrt{n}$, n needs to be quadrupled: $s_Y \div \sqrt{200} = s_Y \div \sqrt{4}\sqrt{50}$ is half of $s_Y \div \sqrt{50}$.
13. D (128)
14. D (127–128)
15. B (125–126)

Chapter 7

1. E (138–142)
2. B (142–143)
3. D (142–143)
4. B (142)
5. A (144–146)
6. C (141)
7. C (138, 142)
8. A (159)
9. C (149–150)
10. B (153–154)
11. A (147–148)
12. C (117) A robust test is analogous to a robust estimator.
13. C (156)
14. D (156–159)
15. A (156–157)

Chapter 8

1. E (148)
2. E (175–176)
3. A (172–174)
4. D (176–178)
5. D (176–178, 182)
6. C (182) $(2 - 1)(4 - 1)$
7. D (176–178, 182)
8. E (182)

9. D. (181–182)
10. A (181–182)
11. B (184–185)
12. C (148)
13. B. (48, 176–178) If the proportions of exposed cases and controls are the same, the odds ration is 1. The 95% confidence interval for the odds ratio does not include 1; this indicates that the proportions are significantly different at the 5% level ($p < 0.05$).
14. B (185–186)
15. E (186–187)

Chapter 9

1. D (205–208)
2. B (203–205, 213)
3. D (214)
4. A (208–209)
5. B (208–209)
6. E (208–209, 214–215)
7. C (208–210)
8. E (198)
9. C (207–208)
10. D (210–211)
11. A (212–213)
12. D (210–213)
13. D (210–212)
14. B (214)
15. E (215–216)

Chapter 10

1. C (241)
2. E (241)
3. C (229–230)
4. A (235)
5. B (235)
6. A (230–232)

7. A (232)
8. E (231, 234) Note that neither of the two other expected mean squares is the same as that for clinics when $\sigma_c^2 = 0$.
9. D (231, 238–239)
10. B (236)
11. E (236–237)
12. E (240)
13. D (240–241) (using the pretreatment value as a covariate)
14. C (240–241)
15. A (240)

Chapter 11

1. A (240–241)
2. E (251–252)
3. B (253–254)
4. C (252)
5. D (254–255)
6. B (258–259)
7. C (257)
8. C (256–257)
9. A (210–211)
10. D (254–255)
11. C (251–252)
12. C (257–258)
13. D (150, 252–253)
14. B (257)
15. C (251–252)

REVIEW PROBLEMS

1. D (4–5)
2. E (4)
3. B (4–5)
4. B (6–7) Contact with a statistician should be initiated as early as possible, but it is not a statistician's role to help decide what is medically appropriate.

5. D (75–77) Use Bayes' theorem: (0.2) (0.95) ÷ [(0.2) (0.95) + (0.8) (0.1)].
6. C (123–124)
7. A (39–40)
8. A (55–57)
9. E (208–209)
10. D (102, 105)
11. B (114)
12. C (141)
13. D (115–116)
14. A (138–142)
15. D (101, 105)
16. E (140–142)
17. A (71) This is the correct answer to the question as worded, but it is not the combined p-value. The combined p-value would be the probability that all three results *or anything more extreme* could have occurred by chance, and is obtained as indicated on page 186.
18. D (155–156)
19. E (185)
20. A (177–178)
21. B (119, 123–126)
22. C (137–138)
23. A (227–229)
24. E (5, 15)
25. D (157)
26. E (3–8)
27. D (143)
28. B (70–71)
29. B (56)
30. D (96) The probability that all three operations are successes is $(3/4)^3 = 27/64$. Therefore, the probability of at least one failure is $1 - 27/64 = 37/64$.
31. E (18)
32. C (21)
33. C (26)
34. A (54)
35. D (46–47)
36. E (215–216)

37. B (16)
38. B (151)
39. C (25)
40. A (46–48)
41. C (22)
42. B (141–142, 159)
43. C (46–47)
44. C (55)
45. B (198–199, 204–209)
46. E (117)
47. A (240)
48. C (240–241)
49. E (253–254)
50. A (138–139, 256–257)

APPENDIX 1

Additional Notes and Computational Formulas

CHAPTER 3

1. The Greek capital sigma, Σ, is the mathematical sign for summation. If we have a sample of n observations, say $y_1, y_2, y_3, \ldots, y_n$, their sum is

$$y_1 + y_2 + y_3 + \cdots + y_n$$

This also can be written $\sum_{i=1}^{n} y_i$, which is read as "the sum of the y_i, for i running from 1 to n." Thus the computational formula for the sample mean, which we shall denote $\bar{y}$, is

$$\bar{y} = \frac{1}{n}\sum_{i=1}^{n} y_i$$

2. Using the same notation, the variance is

$$s^2 = \frac{1}{n-1}\sum_{i=1}^{n} (y_i - \bar{y})^2$$

$$= \frac{1}{n-1}\sum_{i=1}^{n} d_i^2 \qquad \text{where } d_i = y_i - \bar{y}$$

This is the best formula to use on a computer that can store all the y_i in memory. If a calculator is being used, it is usually more convenient to use the mathematically equivalent formula

$$s^2 = \frac{1}{n-1}\left[\sum_{i=1}^{n} y_i^2 - \frac{1}{n}\left(\sum_{i=1}^{n} y_i\right)^2\right]$$

First $\sum_{i=1}^{n} y_i$ is calculated, squared, and divided by n; this gives the "correction factor" $\frac{1}{n}\left(\sum_{i=1}^{n} y_i\right)^2$. Then $\sum_{i=1}^{n} y_i^2$ is calculated and the "correction factor" subtracted from it. The result is divided by $n - 1$.

3. The coefficients of skewness and kurtosis are, respectively,

$$\frac{n^{1/2}\sum_{i=1}^{n}(y_i - \bar{y})^3}{\left[\sum_{i=1}^{n}(y_i - \bar{y})^2\right]^{3/2}} \quad \text{and} \quad \frac{n\sum_{i=1}^{n}(y_i - \bar{y})^4}{\left[\sum_{i=1}^{n}(y_i - \bar{y})^2\right]^{2}}$$

CHAPTER 4

1. Bayes' theorem can be derived quite simply from the basic laws of probability. We have

$$P(D_j|S) = \frac{P(D_j \text{ and } S)}{P(S)}$$

$$\begin{aligned}\text{and } P(S) &= P(D_1 \text{ and } S, \text{ or } D_2 \text{ and } S, \ldots, \text{ or } D_k \text{ and } S)\\ &= P(D_1 \text{ and } S) + P(D_2 \text{ and } S) + \cdots + P(D_k \text{ and } S)\end{aligned}$$

because $D_1, D_2, \ldots, D_k$ are mutually exclusive. It follows immediately that

$$P(D_j|S) = \frac{P(D_j \text{ and } S)}{P(D_1 \text{ and } S) + P(D_2 \text{ and } S) + \cdots + P(D_k \text{ and } S)}$$

2. Data collected by a paternity testing laboratory can be used to estimate the proportion of past cases in which the alleged father was the true father, without any knowledge of whether in each case the alleged father was or was not the true father. Briefly, one method of doing this is as follows. From a large series of cases that have come to the laboratory we can calculate the proportion of cases that resulted in an exclusion, which we shall call P(E). Using genetic theory and a knowledge of the gene frequencies in the population, we can calculate $P(E \mid D_2)$, the probability of exclusion given that a random man is the

true father [see, for example, MacCluer and Schull (1963) Am. J. Hum. Genet. 15:191]. If we assume only two mutually exclusive possibilities exist—D_1, the alleged father is the true father, or D_2, a random man is the true father—we must have

$$\begin{aligned} P(E) &= P(D_1 \text{ and } E) + P(D_2 \text{ and } E) \\ &= P(D_1)P(E \mid D_1) + P(D_2)P(E \mid D_2) \end{aligned}$$

But $P(E \mid D_1)$—the probability of exclusion given that the alleged father is the true father—is zero (whenever the alleged father is the true father, he will not be excluded by the blood testing). Therefore,

$$P(E) = P(D_2)P(E \mid D_2)$$

and so we can calculate $P(D_2) = P(E)/P(E \mid D_2)$, and hence $P(D_1) = 1 - P(D_2)$. In situations in which this "prior probability of paternity" has been estimated, it has been found to be typically between 0.65 and 0.9.

CHAPTER 6

1. There is a subtle difference between the "probability" of a particular sample and its "likelihood." We use the word likelihood when we want to stress that we are interested in different values of the parameters, keeping the values of the random variables fixed at the values observed in a particular sample. We use the word probability, on the other hand, when we want to stress that we are interested in different values of the random variable, keeping the values of the parameters fixed at the population values. Thus the expression for P(y males) on page 117 is both the probability and the likelihood of the sample, which term we use merely indicating whether we want to consider it as a mathematical function of y or of π. Also, using the word "likelihood" avoids the awkward fact that when we are dealing with continuous random variables, any particular sample has (theoretically) zero probability of occurring. The likelihood of a particular y in this case is the height of the probability density function at the point y, but, as explained in Chapter 5, that height is not a probability.

2. The asymptotic properties of unbiasedness, efficiency, and normality only hold in general for maximum likelihood estimators if the likelihood is made a maximum in the mathematical sense (i.e., when the estimates are substituted for the parameters, the likelihood has to be larger than when any other neighboring values are substituted). A likelihood, just like a probability density function, can have one mode, several modes, or no modes; each modal value is a particular maximum likelihood estimate of the parameter. Thus, maximum likelihood estimates may not be unique, nor need they even exist for permissible values of the parameters. A variance, for example, must be positive, and yet its maximum likelihood estimate may sometimes be negative. In a situation such as this the maximum likelihood estimate is sometimes said to be zero, if at zero the likelihood is largest for all permissible values of the parameter. In genetics, the recombination fraction cannot be greater than 0.5. For this reason the maximum likelihood estimate is often said to be 0.5 if the true maximum occurs at some value greater than 0.5. Since in these cases the likelihood is not, however, at a mathematical maximum (i.e., it is not at a mode), such estimators do not possess the same good properties, asymptotically, that true maximum likelihood estimators possess.

3. Even if a maximum likelihood estimate is obtained as a mathematical maximum of the likelihood, there are still certain cases in which, however large the sample, it is still not unbiased. This occurs if, as the sample becomes larger and larger, so does the number of unknown parameters. In such situations the problem is often overcome by maximizing a so-called conditional likelihood. You are most likely to see this term in connection with matched-pair designs. You may also encounter the term "partial likelihood"; estimators that maximize this have the same usual asymptotic properties of maximum likelihood estimators.

4. To show the equivalence between

$$P(-2 \leq \frac{\bar{Y} - \mu}{\sigma_{\bar{Y}}} \leq 2) \doteq 0.95$$

and

$$P(\bar{Y} - 2\sigma_{\bar{Y}} \leq \mu \leq \bar{Y} + 2\sigma_{\bar{Y}}) \doteq 0.95$$

first note that within each probability statement there are two inequalities. We can manipulate these inequalities by the ordinary rules of algebra, without changing the probabilities, as follows (read the sign

<=> as "is equivalent to"):

$$-2 \leq \frac{\overline{Y} - \mu}{\sigma_{\overline{Y}}} \leq 2$$

$$<=> -2 \leq \frac{\overline{Y} - \mu}{\sigma_{\overline{Y}}} \text{ and } \frac{\overline{Y} - \mu}{\sigma_{\overline{Y}}} \leq 2$$

$$<=> -2\sigma_{\overline{Y}} \leq \overline{Y} - \mu \text{ and } \overline{Y} - \mu \leq 2\sigma_{\overline{Y}}$$

$$<=> \mu - 2\sigma_{\overline{Y}} \leq \overline{Y} \text{ and } \overline{Y} \leq \mu + 2\sigma_{\overline{Y}}$$

$$<=> \mu \leq \overline{Y} + 2\sigma_{\overline{Y}} \text{ and } \overline{Y} - 2\sigma_{\overline{Y}} \leq \mu$$

$$<=> \overline{Y} - 2\sigma_{\overline{Y}} \leq \mu \leq \overline{Y} + 2\sigma_{\overline{Y}}$$

CHAPTER 7

1. The null hypothesis must be very specific, since we need to know the distribution of the test criterion under it. If our research hypothesis is, for example, $\pi < 0.6$, we want to disprove the alternative $\pi \geq 0.6$. But, to be specific, we take as our null hypothesis $\pi = 0.6$. Clearly, if we reject the hypothesis $\pi = 0.6$ in favor of $\pi < 0.6$, we must reject with even greater conviction the possibility that $\pi > 0.6$. In other words, whenever an inequality is involved in what we try to disprove, we take as our specific null hypothesis the equality that is closest to the research hypothesis; for if we disprove that particular equality (e.g., $\pi = 0.6$), we shall have automatically disproved the whole inequality (e.g., $\pi \geq 0.6$).

2. The mean of the statistic T for the rank sum test can be derived as follows. First, note that the average of N numbers 1, 2, . . . , N is (N + 1)/2. Under the null hypothesis, T is the sum of n_1 randomly picked numbers from the set of numbers 1, 2, . . . , $n_1 + n_2$. So, putting $N = n_1 + n_2$, their average is $(n_1 + n_2 + 1)/2$; therefore, the sum of n_1 random numbers from the set would be expected, on an average, to be $n_1(n_1 + n_2 + 1)/2$. If T is less than this we would suspect that the median of the first population is less than that of the second, and conversely if T is greater than this. It is not so simple to derive the standard deviation of T.

3. At the end of the chapter we give an expression, using Bayes' theorem, for the posterior probability that a null hypothesis H_0 is false once it

has been rejected by a test of hypothesis. It is assumed in this expression that the alternative to "H_0 is true" is "H_0 is false." Usually, however, there is not just a single alternative, "H_0 is false," but rather a range of alternatives, such as all the different values that the difference between two means, $\mu_1 - \mu_2$, can take on, as illustrated in Figure 7–2. Thus we need to consider, for this example, the prior probability of each different value of $\mu_1 - \mu_2$, multiply it by P (reject $H_0 \mid \mu_1 - \mu_2$), and then sum over all the products for which $\mu_1 - \mu_2 \neq 0$. The result of doing this then replaces P(H_0 is false) P(reject $H_0 \mid H_0$ is false) in both the numerator and denominator of the expression given. Because $\mu_1 - \mu_2$, considered as a variable, is continuous, the prior probabilities can be expressed as a density and integration would then replace the summation. Furthermore, it is doubtful whether we would ever be in a situation where the equality $\mu_1 = \mu_2$ holds exactly, and we would probably be more interested in knowing the probability that $|\mu_1 - \mu_2| \geq \delta$, where δ is the smallest value of the mean difference that is *clinically* significant. Thus the term that is in both the numerator and denominator would be integrated over the values $|\mu_1 - \mu_2| \geq \delta$, and the second term in the denominator would be an analogous integral over the values $|\mu_1 - \mu_2| < \delta$.

CHAPTER 8

1. For any quantity x^2, we have

$$x^2 = (1 - \pi_1)x^2 + \pi_1 x^2$$

Thus

$$\frac{(y_1 - n\pi_1)^2}{n\pi_1(1 - \pi_1)} = (1 - \pi_1)\frac{(y_1 - n\pi_1)^2}{n\pi_1(1 - \pi_1)} + \pi_1\frac{(y_1 - n\pi_1)^2}{n\pi_1(1 - \pi_1)}$$

$$= \frac{(y_1 - n\pi_1)^2}{n\pi_1} + \frac{(y_1 - n\pi_1)^2}{n(1 - \pi_1)}$$

Now substitute $y_1 = n - y_2$ and $\pi_1 = 1 - \pi_2$ in the numerator of the second term; we obtain

$$\frac{(y_1 - n\pi_1)^2}{n\pi_1} + \frac{[(n - y_2) - n(1 - \pi_2)]^2}{n(1 - \pi_1)} = \frac{(y_1 - n\pi_1)^2}{n\pi_1} + \frac{(-y_2 + n\pi_2)^2}{n\pi_2}$$

$$= \frac{(y_1 - n\pi_1)^2}{n\pi_1} + \frac{(y_2 - n\pi_2)^2}{n\pi_2}$$

2. There is another (mathematically identical) formula for calculating the Pearson chi-square statistic from a 2 × 2 contingency table. Suppose we write the table as follows:

a	b	(a + b)
c	d	(c + d)
(a + c)	(b + d)	N = a + b + c + d

The formula is then

$$\frac{(ad - bc)^2 N}{(a + b)(c + d)(a + c)(b + d)}$$

Sometimes this formula is modified to include a so-called correction for continuity that makes the resulting chi-square smaller. Provided each expected value is at least 5, however, there is no need for this modification.

3. Suppose we calculate the usual contingency table chi-square to test for independence between two dependent traits when we have matched pairs. As in Chapter 8, suppose each pair consists of a man and a woman matched for age, and the dependent variables are cholesterol level of the man and cholesterol level of the woman in each pair. Then the usual contingency table chi-square would be relevant for answering the question: is the cholesterol level of the woman in the pair independent of the cholesterol level of the man in the pair? (i.e., Does age, the matching variable, have a common effect on the cholesterol level of both men and women?) In other words, in this type of study the usual contingency table chi-square tests whether the matching was necessary—a significant result indicating that it was, a nonsignificant result indicating it was not.

4. The likelihood ratio criterion is a general criterion for testing one hypothesis against any other. Often we test a specific null hypothesis under a general statistical model: the null hypothesis is a special case of the model. In the example in Chapter 8, the model for the two samples of data is that they come from two normal distributions, both of which have the same variance. Under this model, the means of the two distributions could be the same or different. We test the null hypothesis that they are the same. In order for $-2 \log_e$ (likelihood ratio) to be asymptotically distributed as chi-square, it is necessary for the null

hypothesis to be a special case of a general model. The numerator of the likelihood ratio is maximized under the null hypothesis, and the denominator is maximized under the model. The likelihood ratio then lies between 0 and 1, and $-2 \log_e$ (the likelihood ratio) cannot be negative.

CHAPTER 9

1. For a sample of n pairs, let $\bar{x}$ and $\bar{y}$ be the sample means. Define the quantities:

$$SS_x = \sum_{i=1}^{n} (x_i - \bar{x})^2 = \sum_{i=1}^{n} x_i^2 - \frac{1}{n}\left(\sum_{i=1}^{n} x_i\right)^2$$

$$SS_y = \sum_{i=1}^{n} (y_i - \bar{y})^2 = \sum_{i=1}^{n} y_i^2 - \frac{1}{n}\left(\sum_{i=1}^{n} y_i\right)^2$$

$$SS_{xy} = \sum_{i=1}^{n} (x_i - \bar{x})(y_i - \bar{y}) = \sum_{i=1}^{n} x_i y_i - \frac{1}{n}\left(\sum_{i=1}^{n} x_i\right)\left(\sum_{i=1}^{n} y_i\right)$$

Thus SS_x is the total sum of the squared deviations of the x_i from the mean $\bar{x}$, and SS_y is the total sum of the squared deviations of the y_i from the mean $\bar{y}$. SS_{xy} is the analogous total sum of the cross-products; and the same considerations govern which formula to use for calculating this as govern the calculation of SS_x and SS_y. (See note 2 for Chapter 3 in this appendix.) Using these quantities, we successively calculate the estimated regression coefficient b_1 and intercept b_0, for the regression line $\hat{y}_i = b_0 + b_1 x_i$, as follows:

$$b_1 = \frac{SS_{xy}}{SS_x}$$

$$b_0 = \bar{y} - \frac{SS_{xy}}{SS_x}\bar{x} = \bar{y} - b_1\bar{x}$$

The sum of squares in Table 9–1 can then be calculated as

$$SS_R = \frac{SS_{xy}^2}{SS_x} = b_1 SS_{xy}$$

$$SS_E = SS_y - \frac{SS_{xy}^2}{SS_x} = SS_y - b_1 SS_{xy}$$

and, as explained, the mean squares are obtained by dividing the sums of the squares by their respective d.f.; i.e.,

$$MS_R = \frac{SS_R}{1} = SS_R$$

$$MS_E = \frac{SS_E}{n - 2}$$

Note that the total sum of squares is

$$SS_T = SS_R + SS_E = SS_y$$

(i.e., the total sum of squares of the dependent variable Y about its mean). The standard error of b_1 is $\sqrt{MS_E/SS_x}$.

2. The estimated regression line of X on y is given by

$$b_1 = \frac{SS_{xy}}{SS_y}$$

$$b_0 = \bar{x} - \frac{SS_{xy}}{SS_y}\bar{y} = \bar{x} - b_1\bar{y}$$

The covariance between X and Y is estimated by $SS_{xy}/(n - 1)$, and the correlation by $SS_{xy}/\sqrt{SS_xSS_y}$.

3. It is instructive to check, in the tables of Appendix 2, that the square of the 97.5th percentile of Student's t with k d.f. is equal to the 95th percentile of F with 1 and k d.f. Similarly, the square of the 95th percentile of t with k d.f. is equal to the 90th percentile of F with 1 and k d.f. Just as χ^2 with 1 d.f. is the square of a normal random variable, so F with 1 and k d.f. is the square of a random variable distributed as t with k d.f. Thus, articles you read in the literature may use either of these equivalent test statistics.

CHAPTER 10

1. It is possible to calculate the mean squares in Table 10–3 from the data presented in Table 10–2. The mean square among drug groups is 10 (the number of patients in each group) times the variance of the four group means (i.e., 10 times the variance of the set of four numbers 80, 94, 92, and 90). The mean square within drug groups is the pooled within-group variance. Since each group contains the same number of patients, this is the simple average of the four variances, i.e.,

$$\frac{(10.5)^2 + (9.5)^2 + (9.7)^2 + (10.2)^2}{4}$$

2. There are several so-called *multiple comparison* procedures that can be used to test which pairs of a set of means are significantly different. The simplest (Fisher's least significant difference method) is to perform Student's t-test on all the pairwise comparisons, but to require in addition, before making the pairwise comparisons, a significant overall F-test for the equality of the means being considered. The t-tests are performed using the pooled estimate of σ^2 given by the error mean square of the analysis of variance. Another procedure (the Fisher-Bonferroni method) similarly uses multiple t-tests, but each test must reach significance at the α/c level, where c is the total number of comparisons made, to ensure an overall significance level of α.

Several procedures (Tukey's, Newman-Keuls', and Duncan's) begin by comparing the largest mean with the smallest, and continue with the next largest difference, and so on, until either a nonsignificant result is encountered or until all pairwise comparisons have been made. At each step the difference is compared to an appropriate null distribution.

A further method (Scheffé's) allows for testing more complex contrasts (such as H_0: $\mu_1 - \mu_2 + \mu_3 - \mu_4 = 0$), as well as all pairwise comparisons of means. Finally, there is a procedure (Dunnett's) aimed specifically at identifying group means that are significantly different from the mean of a control group.

All these tests are aimed at maintaining the overall significance level at some fixed value, α (i.e., ensuring that, if the null hypothesis is true, the probability of one or more significant differences being found is α).

3. We now give simple rules for writing down formulas for sums of squares that apply in the case of any balanced design (i.e., any design in which all groups of a particular type are the same size). The rules are easy to apply if the degrees of freedom are known for each source of variation. Consider first the "among groups" sums of squares in Table 10–1.

1. The degrees of freedom are $a - 1$.

2. Write down a pair of parentheses and a square (superscript 2) for each term in the degrees of freedom: $(\quad)^2\ (\quad)^2$.

3. In front of each pair of parentheses write the appropriate sign corresponding to the term in the degrees of freedom, and a summation sign for each letter in the degrees of freedom:

$$\sum_{i=1}^{a}(\quad)^2 - (\quad)^2$$

4. Put summation signs inside the parentheses for all factors and replicates that are not summed over outside the parentheses, followed by the symbol for a typical observation:

$$\sum_{i=1}^{a}\left(\sum_{k=1}^{n} y_{ik}\right)^2 - \left(\sum_{i=1}^{a}\sum_{k=1}^{n} y_{ik}\right)^2$$

5. Finally, divide each term by the number of observations summed within the parentheses. Thus

$$SS_A = \frac{\sum_{i=1}^{a}\left(\sum_{k=1}^{n} y_{ik}\right)^2}{n} - \frac{\left(\sum_{i=1}^{a}\sum_{k=1}^{n} y_{ik}\right)^2}{an}$$

Applying this same sequence of rules, we obtain the sum of squares within groups for Table 10–1, noting that we multiply out the expression for the degrees of freedom.

1. $an - a$

2. $(\quad)^2\ (\quad)^2$

3. $\sum_{i=1}^{a}\sum_{k=1}^{n}(\quad)^2 - \sum_{i=1}^{a}(\quad)^2$

4. $\sum_{i=1}^{a}\sum_{k=1}^{n}(y_{ik})^2 - \sum_{i=1}^{a}\left(\sum_{k=1}^{n} y_{ik}\right)^2$

5. $SS_R = \dfrac{\sum_{i=1}^{a}\sum_{k=1}^{n}(y_{ik})^2}{1} - \dfrac{\sum_{i=1}^{a}\left(\sum_{k=1}^{n} y_{ik}\right)^2}{n}$

Similarly, for the nested factor analysis of variance in Table 10–4 we obtain

$$SS_A = \frac{\sum_{i=1}^{a}\left(\sum_{j=1}^{b}\sum_{k=1}^{n} y_{ijk}\right)^2}{bn} - \frac{\left(\sum_{i=1}^{a}\sum_{j=1}^{b}\sum_{k=1}^{n} y_{ijk}\right)^2}{abn}$$

$$SS_B = \frac{\sum_{i=1}^{a}\sum_{j=1}^{b}\left(\sum_{k=1}^{n} y_{ijk}\right)^2}{n} - \frac{\sum_{i=1}^{a}\left(\sum_{j=1}^{b}\sum_{k=1}^{n} y_{ijk}\right)^2}{bn}$$

$$SS_R = \sum_{i=1}^{a}\sum_{j=1}^{b}\sum_{k=1}^{n} y_{ijk}^2 - \frac{\sum_{i=1}^{a}\sum_{j=1}^{b}\left(\sum_{k=1}^{n} y_{ijk}\right)^2}{n}$$

For the two-way analysis of variance in Table 10–6 we obtain

$$SS_A = \frac{\sum_{i=1}^{a}\left(\sum_{j=1}^{b}\sum_{k=1}^{n} y_{ijk}\right)^2}{bn} - \frac{\left(\sum_{i=1}^{a}\sum_{j=1}^{b}\sum_{k=1}^{n} y_{ijk}\right)^2}{abn}$$

$$SS_B = \frac{\sum_{j=1}^{b}\left(\sum_{i=1}^{a}\sum_{k=1}^{n} y_{ijk}\right)^2}{an} - \frac{\left(\sum_{i=1}^{a}\sum_{j=1}^{b}\sum_{k=1}^{n} y_{ijk}\right)^2}{abn}$$

$$SS_{AB} = \frac{\sum_{i=1}^{a}\sum_{j=1}^{b}\left(\sum_{k=1}^{n} y_{ijk}\right)^2}{n} - \frac{\sum_{i=1}^{a}\left(\sum_{j=1}^{b}\sum_{k=1}^{n} y_{ijk}\right)^2}{bn} - \frac{\sum_{j=1}^{b}\left(\sum_{i=1}^{a}\sum_{k=1}^{n} y_{ijk}\right)^2}{an} + \frac{\left(\sum_{i=1}^{a}\sum_{j=1}^{b}\sum_{k=1}^{n} y_{ijk}\right)^2}{abn}$$

$$SS_R = \sum_{i=1}^{a}\sum_{j=1}^{b}\sum_{k=1}^{n} y_{ijk}^2 - \frac{\sum_{i=1}^{a}\sum_{j=1}^{b}\left(\sum_{k=1}^{n} y_{ijk}\right)^2}{n}$$

APPENDIX 2

Statistical Tables

TABLE A2-1. **Standard Normal Distribution (P(Z ≤ z)**

z	.00	.01	.02	.03	.04	.05	.06	.07	.08	.09
−3.4	.0003	.0003	.0003	.0003	.0003	.0003	.0003	.0003	.0003	.0002
−3.3	.0005	.0005	.0005	.0004	.0004	.0004	.0004	.0004	.0004	.0003
−3.2	.0007	.0007	.0006	.0006	.0006	.0006	.0006	.0005	.0005	.0005
−3.1	.0010	.0009	.0009	.0009	.0008	.0008	.0008	.0008	.0007	.0007
−3.0	.0013	.0013	.0013	.0012	.0012	.0011	.0011	.0011	.0010	.0010
−2.9	.0019	.0018	.0018	.0017	.0016	.0016	.0015	.0015	.0014	.0014
−2.8	.0026	.0025	.0024	.0023	.0023	.0022	.0021	.0021	.0020	.0019
−2.7	.0035	.0034	.0033	.0032	.0031	.0030	.0029	.0028	.0027	.0026
−2.6	.0047	.0045	.0044	.0043	.0041	.0040	.0039	.0038	.0037	.0036
−2.5	.0062	.0060	.0059	.0057	.0055	.0054	.0052	.0051	.0049	.0048
−2.4	.0082	.0080	.0078	.0075	.0073	.0071	.0069	.0068	.0066	.0064
−2.3	.0107	.0104	.0102	.0099	.0096	.0094	.0091	.0089	.0087	.0084
−2.2	.0139	.0136	.0132	.0129	.0125	.0122	.0119	.0116	.0113	.0110
−2.1	.0179	.0174	.0170	.0166	.0162	.0158	.0154	.0150	.0146	.0143
−2.0	.0228	.0222	.0217	.0212	.0207	.0202	.0197	.0192	.0188	.0183
−1.9	.0287	.0281	.0274	.0268	.0262	.0256	.0250	.0244	.0239	.0233
−1.8	.0359	.0351	.0344	.0336	.0329	.0322	.0314	.0307	.0301	.0294
−1.7	.0446	.0436	.0427	.0418	.0409	.0401	.0392	.0384	.0375	.0367
−1.6	.0548	.0537	.0526	.0516	.0505	.0495	.0485	.0475	.0465	.0455
−1.5	.0668	.0655	.0643	.0630	.0618	.0606	.0594	.0582	.0571	.0559
−1.4	.0808	.0793	.0778	.0764	.0749	.0735	.0721	.0708	.0694	.0681
−1.3	.0968	.0951	.0934	.0918	.0901	.0885	.0869	.0853	.0838	.0823
−1.2	.1151	.1131	.1112	.1093	.1075	.1056	.1038	.1020	.1003	.0985
−1.1	.1357	.1335	.1314	.1292	.1271	.1251	.1230	.1210	.1190	.1170
−1.0	.1587	.1562	.1539	.1515	.1492	.1469	.1446	.1423	.1401	.1379
−0.9	.1841	.1814	.1788	.1762	.1736	.1711	.1685	.1660	.1635	.1611
−0.8	.2119	.2090	.2061	.2033	.2005	.1977	.1949	.1922	.1894	.1867
−0.7	.2420	.2389	.2358	.2327	.2296	.2266	.2236	.2206	.2177	.2148
−0.6	.2743	.2709	.2676	.2643	.2611	.2578	.2546	.2514	.2483	.2451
−0.5	.3085	.3050	.3015	.2981	.2946	.2912	.2877	.2843	.2810	.2776
−0.4	.3446	.3409	.3372	.3336	.3300	.3264	.3228	.3192	.3156	.3121
−0.3	.3821	.3783	.3745	.3707	.3669	.3632	.3594	.3557	.3520	.3483
−0.2	.4207	.4168	.4129	.4090	.4052	.4013	.3974	.3936	.3897	.3859
−0.1	.4602	.4562	.4522	.4483	.4443	.4404	.4364	.4325	.4286	.4247

TABLE A2-1. **Standard Normal Distribution (P(Z ≤ z) (*Continued*)**

z	.00	.01	.02	.03	.04	.05	.06	.07	.08	.09
−0.0	.5000	.4960	.4920	.4880	.4840	.4801	.4761	.4721	.4681	.4641
0.0	.5000	.5040	.5080	.5120	.5160	.5199	.5239	.5279	.5319	.5359
0.1	.5398	.5438	.5478	.5517	.5557	.5596	.5636	.5675	.5714	.5753
0.2	.5793	.5832	.5871	.5910	.5948	.5987	.6026	.6064	.6103	.6141
0.3	.6179	.6217	.6255	.6293	.6331	.6368	.6406	.6443	.6480	.6517
0.4	.6554	.6591	.6628	.6664	.6700	.6736	.6772	.6808	.6844	.6879
0.5	.6915	.6950	.6985	.7019	.7054	.7088	.7123	.7157	.7190	.7224
0.6	.7257	.7291	.7324	.7357	.7389	.7422	.7454	.7486	.7517	.7549
0.7	.7580	.7611	.7642	.7673	.7704	.7734	.7764	.7794	.7823	.7852
0.8	.7881	.7910	.7939	.7967	.7995	.8023	.8051	.8078	.8106	.8133
0.9	.8159	.8186	.8212	.8238	.8264	.8289	.8315	.8340	.8365	.8389
1.0	.8413	.8438	.8461	.8485	.8508	.8531	.8554	.8577	.8599	.8621
1.1	.8643	.8665	.8686	.8708	.8729	.8749	.8770	.8790	.8810	.8830
1.2	.8849	.8869	.8888	.8907	.8925	.8944	.8962	.8980	.8997	.9015
1.3	.9032	.9049	.9066	.9082	.9099	.9115	.9131	.9147	.9162	.9177
1.4	.9192	.9207	.9222	.9236	.9251	.9265	.9279	.9292	.9306	.9319
1.5	.9332	.9345	.9357	.9370	.9382	.9394	.9406	.9418	.9429	.9441
1.6	.9452	.9463	.9474	.9484	.9495	.9505	.9515	.9525	.9535	.9545
1.7	.9554	.9564	.9573	.9582	.9591	.9599	.9608	.9616	.9625	.9633
1.8	.9641	.9649	.9656	.9664	.9671	.9678	.9686	.9693	.9699	.9706
1.9	.9713	.9719	.9726	.9732	.9738	.9744	.9750	.9756	.9761	.9767
2.0	.9772	.9778	.9783	.9788	.9793	.9798	.9803	.9808	.9812	.9817
2.1	.9821	.9826	.9830	.9834	.9838	.9842	.9846	.9850	.9854	.9857
2.2	.9861	.9864	.9868	.9871	.9875	.9878	.9881	.9884	.9887	.9890
2.3	.9893	.9896	.9898	.9901	.9904	.9906	.9909	.9911	.9913	.9916
2.4	.9918	.9920	.9922	.9925	.9927	.9929	.9931	.9932	.9934	.9936
2.5	.9938	.9940	.9941	.9943	.9945	.9946	.9948	.9949	.9951	.9952
2.6	.9953	.9955	.9956	.9957	.9959	.9960	.9961	.9962	.9963	.9964
2.7	.9965	.9966	.9967	.9968	.9969	.9970	.9971	.9972	.9973	.9974
2.8	.9974	.9975	.9976	.9977	.9977	.9978	.9979	.9979	.9980	.9981
2.9	.9981	.9982	.9982	.9983	.9984	.9984	.9985	.9985	.9986	.9986
3.0	.9987	.9987	.9987	.9988	.9988	.9989	.9989	.9989	.9990	.9990
3.1	.9990	.9991	.9991	.9991	.9992	.9992	.9992	.9992	.9993	.9993
3.2	.9993	.9993	.9994	.9994	.9994	.9994	.9994	.9995	.9995	.9995
3.3	.9995	.9995	.9995	.9996	.9996	.9996	.9996	.9996	.9996	.9997
3.4	.9997	.9997	.9997	.9997	.9997	.9997	.9997	.9997	.9997	.9998

TABLE A2-2. **Percentiles of Student's t-Distribution**

d.f.	t_{60} $-t_{40}$	t_{70} $-t_{30}$	t_{80} $-t_{20}$	t_{90} $-t_{10}$	t_{95} $-t_{5}$	$t_{97.5}$ $-t_{2.5}$	t_{99} $-t_{1}$	$t_{99.5}$ $-t_{0.5}$
1	.325	.727	1.376	3.078	6.314	12.706	31.821	63.657
2	.289	.617	1.061	1.886	2.920	4.303	6.965	9.925
3	.277	.584	.978	1.638	2.353	3.182	4.541	5.841
4	.271	.569	.941	1.533	2.132	2.776	3.747	4.604
5	.267	.559	.920	1.476	2.015	2.571	3.365	4.032
6	.265	.553	.906	1.440	1.943	2.447	3.143	3.707
7	.263	.549	.896	1.415	1.895	2.365	2.998	3.499
8	.262	.546	.889	1.397	1.860	2.306	2.896	3.355
9	.261	.543	.883	1.383	1.833	2.262	2.821	3.250
10	.260	.542	.879	1.372	1.812	2.228	2.764	3.169
11	.260	.540	.876	1.363	1.796	2.201	2.718	3.106
12	.259	.539	.873	1.356	1.782	2.179	2.681	3.055
13	.259	.538	.870	1.350	1.771	2.160	2.650	3.012
14	.258	.537	.868	1.345	1.761	2.145	2.624	2.977
15	.258	.536	.866	1.341	1.753	2.131	2.602	2.947
16	.258	.535	.865	1.337	1.746	2.120	2.583	2.921
17	.257	.534	.863	1.333	1.740	2.110	2.567	2.898
18	.257	.534	.862	1.330	1.734	2.101	2.552	2.878
19	.257	.533	.861	1.328	1.729	2.093	2.539	2.861
20	.257	.533	.860	1.325	1.725	2.086	2.528	2.845
21	.257	.532	.859	1.323	1.721	2.080	2.518	2.831
22	.256	.532	.858	1.321	1.717	2.074	2.508	2.819
23	.256	.532	.858	1.319	1.714	2.069	2.500	2.807
24	.256	.531	.857	1.318	1.711	2.064	2.492	2.797
25	.256	.531	.856	1.316	1.708	2.060	2.485	2.787
26	.256	.531	.856	1.315	1.706	2.056	2.479	2.779
27	.256	.531	.855	1.314	1.703	2.052	2.473	2.771
28	.256	.530	.855	1.313	1.701	2.048	2.467	2.763
29	.256	.530	.854	1.311	1.699	2.045	2.462	2.756
30	.256	.530	.854	1.310	1.697	2.042	2.457	2.750
40	.255	.529	.851	1.303	1.684	2.021	2.423	2.704
60	.254	.527	.848	1.296	1.671	2.000	2.390	2.660
120	.254	.526	.845	1.289	1.658	1.980	2.358	2.617
∞	.253	.524	.842	1.282	1.645	1.960	2.326	2.576

Source: Adapted from Table III of Fisher and Yates: *Statistical Tables for Biological, Agricultural and Medical Research,* Ed. 6. Hafner Publishing Co., 1963, with permission.

TABLE A2-3. **Percentage Points of the F-Distribution** **90th Percentiles**

n_2 \ n_1	1	2	3	4	5	6	7	8	9	10	12	15	20	24	30	40	60	120	∞
1	647·8	799·5	864·2	899·6	921·8	937·1	948·2	956·7	963·3	968·6	976·7	984·9	993·1	997·2	1001	1006	1010	1014	1018
2	38·51	39·00	39·17	39·25	39·30	39·33	39·36	39·37	39·39	39·40	39·41	39·43	39·45	39·46	39·46	39·47	39·48	39·49	39·50
3	17·44	16·04	15·44	15·10	14·88	14·73	14·62	14·54	14·47	14·42	14·34	14·25	14·17	14·12	14·08	14·04	13·99	13·95	13·90
4	12·22	10·65	9·98	9·60	9·36	9·20	9·07	8·98	8·90	8·84	8·75	8·66	8·56	8·51	8·46	8·41	8·36	8·31	8·26
5	10·01	8·43	7·76	7·39	7·15	6·98	6·85	6·76	6·68	6·62	6·52	6·43	6·33	6·28	6·23	6·18	6·12	6·07	6·02
6	8·81	7·26	6·60	6·23	5·99	5·82	5·70	5·60	5·52	5·46	5·37	5·27	5·17	5·12	5·07	5·01	4·96	4·90	4·85
7	8·07	6·54	5·89	5·52	5·29	5·12	4·99	4·90	4·82	4·76	4·67	4·57	4·47	4·42	4·36	4·31	4·25	4·20	4·14
8	7·57	6·06	5·42	5·05	4·82	4·65	4·53	4·43	4·36	4·30	4·20	4·10	4·00	3·95	3·89	3·84	3·78	3·73	3·67
9	7·21	5·71	5·08	4·72	4·48	4·32	4·20	4·10	4·03	3·96	3·87	3·77	3·67	3·61	3·56	3·51	3·45	3·39	3·33
10	6·94	5·46	4·83	4·47	4·24	4·07	3·95	3·85	3·78	3·72	3·62	3·52	3·42	3·37	3·31	3·26	3·20	3·14	3·08
11	6·72	5·26	4·63	4·28	4·04	3·88	3·76	3·66	3·59	3·53	3·43	3·33	3·23	3·17	3·12	3·06	3·00	2·94	2·88
12	6·55	5·10	4·47	4·12	3·89	3·73	3·61	3·51	3·44	3·37	3·28	3·18	3·07	3·02	2·96	2·91	2·85	2·79	2·72
13	6·41	4·97	4·35	4·00	3·77	3·60	3·48	3·39	3·31	3·25	3·15	3·05	2·95	2·89	2·84	2·78	2·72	2·66	2·60
14	6·30	4·86	4·24	3·89	3·66	3·50	3·38	3·29	3·21	3·15	3·05	2·95	2·84	2·79	2·73	2·67	2·61	2·55	2·49
15	6·20	4·77	4·15	3·80	3·58	3·41	3·29	3·20	3·12	3·06	2·96	2·86	2·76	2·70	2·64	2·59	2·52	2·46	2·40
16	6·12	4·69	4·08	3·73	3·50	3·34	3·22	3·12	3·05	2·99	2·89	2·79	2·68	2·63	2·57	2·51	2·45	2·38	2·32
17	6·04	4·62	4·01	3·66	3·44	3·28	3·16	3·06	2·98	2·92	2·82	2·72	2·62	2·56	2·50	2·44	2·38	2·32	2·25
18	5·98	4·56	3·95	3·61	3·38	3·22	3·10	3·01	2·93	2·87	2·77	2·67	2·56	2·50	2·44	2·38	2·32	2·26	2·19
19	5·92	4·51	3·90	3·56	3·33	3·17	3·05	2·96	2·88	2·82	2·72	2·62	2·51	2·45	2·39	2·33	2·27	2·20	2·13
20	5·87	4·46	3·86	3·51	3·29	3·13	3·01	2·91	2·84	2·77	2·68	2·57	2·46	2·41	2·35	2·29	2·22	2·16	2·09
21	5·83	4·42	3·82	3·48	3·25	3·09	2·97	2·87	2·80	2·73	2·64	2·53	2·42	2·37	2·31	2·25	2·18	2·11	2·04
22	5·79	4·38	3·78	3·44	3·22	3·05	2·93	2·84	2·76	2·70	2·60	2·50	2·39	2·33	2·27	2·21	2·14	2·08	2·00
23	5·75	4·35	3·75	3·41	3·18	3·02	2·90	2·81	2·73	2·67	2·57	2·47	2·36	2·30	2·24	2·18	2·11	2·04	1·97
24	5·72	4·32	3·72	3·38	3·15	2·99	2·87	2·78	2·70	2·64	2·54	2·44	2·33	2·27	2·21	2·15	2·08	2·01	1·94
25	5·69	4·29	3·69	3·35	3·13	2·97	2·85	2·75	2·68	2·61	2·51	2·41	2·30	2·24	2·18	2·12	2·05	1·98	1·91
26	5·66	4·27	3·67	3·33	3·10	2·94	2·82	2·73	2·65	2·59	2·49	2·39	2·28	2·22	2·16	2·09	2·03	1·95	1·88
27	5·63	4·24	3·65	3·31	3·08	2·92	2·80	2·71	2·63	2·57	2·47	2·36	2·25	2·19	2·13	2·07	2·00	1·93	1·85
28	5·61	4·22	3·63	3·29	3·06	2·90	2·78	2·69	2·61	2·55	2·45	2·34	2·23	2·17	2·11	2·05	1·98	1·91	1·83
29	5·59	4·20	3·61	3·27	3·04	2·88	2·76	2·67	2·59	2·53	2·43	2·32	2·21	2·15	2·09	2·03	1·96	1·89	1·81
30	5·57	4·18	3·59	3·25	3·03	2·87	2·75	2·65	2·57	2·51	2·41	2·31	2·20	2·14	2·07	2·01	1·94	1·87	1·79
40	5·42	4·05	3·46	3·13	2·90	2·74	2·62	2·53	2·45	2·39	2·29	2·18	2·07	2·01	1·94	1·88	1·80	1·72	1·64
60	5·29	3·93	3·34	3·01	2·79	2·63	2·51	2·41	2·33	2·27	2·17	2·06	1·94	1·88	1·82	1·74	1·67	1·58	1·48
120	5·15	3·80	3·23	2·89	2·67	2·52	2·39	2·30	2·22	2·16	2·05	1·94	1·82	1·76	1·69	1·61	1·53	1·43	1·31
∞	5·02	3·69	3·12	2·79	2·57	2·41	2·29	2·19	2·11	2·05	1·94	1·83	1·71	1·64	1·57	1·48	1·39	1·27	1·00

95th Percentiles

n_2 \ n_1	1	2	3	4	5	6	7	8	9	10	12	15	20	24	30	40	60	120	∞
1	4052	4999·5	5403	5625	5764	5859	5928	5982	6022	6056	6106	6157	6209	6235	6261	6287	6313	6339	6366
2	98·50	99·00	99·17	99·25	99·30	99·33	99·36	99·37	99·39	99·40	99·42	99·43	99·45	99·46	99·47	99·47	99·48	99·49	99·50
3	34·12	30·82	29·46	28·71	28·24	27·91	27·67	27·49	27·35	27·23	27·05	26·87	26·69	26·60	26·50	26·41	26·32	26·22	26·13
4	21·20	18·00	16·69	15·98	15·52	15·21	14·98	14·80	14·66	14·55	14·37	14·20	14·02	13·93	13·84	13·75	13·65	13·56	13·46
5	16·26	13·27	12·06	11·39	10·97	10·67	10·46	10·29	10·16	10·05	9·89	9·72	9·55	9·47	9·38	9·29	9·20	9·11	9·02
6	13·75	10·92	9·78	9·15	8·75	8·47	8·26	8·10	7·98	7·87	7·72	7·56	7·40	7·31	7·23	7·14	7·06	6·97	6·88
7	12·25	9·55	8·45	7·85	7·46	7·19	6·99	6·84	6·72	6·62	6·47	6·31	6·16	6·07	5·99	5·91	5·82	5·74	5·65
8	11·26	8·65	7·59	7·01	6·63	6·37	6·18	6·03	5·91	5·81	5·67	5·52	5·36	5·28	5·20	5·12	5·03	4·95	4·86
9	10·56	8·02	6·99	6·42	6·06	5·80	5·61	5·47	5·35	5·26	5·11	4·96	4·81	4·73	4·65	4·57	4·48	4·40	4·31
10	10·04	7·56	6·55	5·99	5·64	5·39	5·20	5·06	4·94	4·85	4·71	4·56	4·41	4·33	4·25	4·17	4·08	4·00	3·91
11	9·65	7·21	6·22	5·67	5·32	5·07	4·89	4·74	4·63	4·54	4·40	4·25	4·10	4·02	3·94	3·86	3·78	3·69	3·60
12	9·33	6·93	5·95	5·41	5·06	4·82	4·64	4·50	4·39	4·30	4·16	4·01	3·86	3·78	3·70	3·62	3·54	3·45	3·36
13	9·07	6·70	5·74	5·21	4·86	4·62	4·44	4·30	4·19	4·10	3·96	3·82	3·66	3·59	3·51	3·43	3·34	3·25	3·17
14	8·86	6·51	5·56	5·04	4·69	4·46	4·28	4·14	4·03	3·94	3·80	3·66	3·51	3·43	3·35	3·27	3·18	3·09	3·00
15	8·68	6·36	5·42	4·89	4·56	4·32	4·14	4·00	3·89	3·80	3·67	3·52	3·37	3·29	3·21	3·13	3·05	2·96	2·87
16	8·53	6·23	5·29	4·77	4·44	4·20	4·03	3·89	3·78	3·69	3·55	3·41	3·26	3·18	3·10	3·02	2·93	2·84	2·75
17	8·40	6·11	5·18	4·67	4·34	4·10	3·93	3·79	3·68	3·59	3·46	3·31	3·16	3·08	3·00	2·92	2·83	2·75	2·65
18	8·29	6·01	5·09	4·58	4·25	4·01	3·84	3·71	3·60	3·51	3·37	3·23	3·08	3·00	2·92	2·84	2·75	2·66	2·57
19	8·18	5·93	5·01	4·50	4·17	3·94	3·77	3·63	3·52	3·43	3·30	3·15	3·00	2·92	2·84	2·76	2·67	2·58	2·49
20	8·10	5·85	4·94	4·43	4·10	3·87	3·70	3·56	3·46	3·37	3·23	3·09	2·94	2·86	2·78	2·69	2·61	2·52	2·42
21	8·02	5·78	4·87	4·37	4·04	3·81	3·64	3·51	3·40	3·31	3·17	3·03	2·88	2·80	2·72	2·64	2·55	2·46	2·36
22	7·95	5·72	4·82	4·31	3·99	3·76	3·59	3·45	3·35	3·26	3·12	2·98	2·83	2·75	2·67	2·58	2·50	2·40	2·31
23	7·88	5·66	4·76	4·26	3·94	3·71	3·54	3·41	3·30	3·21	3·07	2·93	2·78	2·70	2·62	2·54	2·45	2·35	2·26
24	7·82	5·61	4·72	4·22	3·90	3·67	3·50	3·36	3·26	3·17	3·03	2·89	2·74	2·66	2·58	2·49	2·40	2·31	2·21
25	7·77	5·57	4·68	4·18	3·85	3·63	3·46	3·32	3·22	3·13	2·99	2·85	2·70	2·62	2·54	2·45	2·36	2·27	2·17
26	7·72	5·53	4·64	4·14	3·82	3·59	3·42	3·29	3·18	3·09	2·96	2·81	2·66	2·58	2·50	2·42	2·33	2·23	2·13
27	7·68	5·49	4·60	4·11	3·78	3·56	3·39	3·26	3·15	3·06	2·93	2·78	2·63	2·55	2·47	2·38	2·29	2·20	2·10
28	7·64	5·45	4·57	4·07	3·75	3·53	3·36	3·23	3·12	3·03	2·90	2·75	2·60	2·52	2·44	2·35	2·26	2·17	2·06
29	7·60	5·42	4·54	4·04	3·73	3·50	3·33	3·20	3·09	3·00	2·87	2·73	2·57	2·49	2·41	2·33	2·23	2·14	2·03
30	7·56	5·39	4·51	4·02	3·70	3·47	3·30	3·17	3·07	2·98	2·84	2·70	2·55	2·47	2·39	2·30	2·21	2·11	2·01
40	7·31	5·18	4·31	3·83	3·51	3·29	3·12	2·99	2·89	2·80	2·66	2·52	2·37	2·29	2·20	2·11	2·02	1·92	1·80
60	7·08	4·98	4·13	3·65	3·34	3·12	2·95	2·82	2·72	2·63	2·50	2·35	2·20	2·12	2·03	1·94	1·84	1·73	1·60
120	6·85	4·79	3·95	3·48	3·17	2·96	2·79	2·66	2·56	2·47	2·34	2·19	2·03	1·95	1·86	1·76	1·66	1·53	1·38
∞	6·63	4·61	3·78	3·32	3·02	2·80	2·64	2·51	2·41	2·32	2·18	2·04	1·88	1·79	1·70	1·59	1·47	1·32	1·00

TABLE A2-3. **Percentage Points of the F-Distribution** (*Continued*) **97.5th Percentiles**

n_2 \ n_1	1	2	3	4	5	6	7	8	9	10	12	15	20	24	30	40	60	120	∞
1	39·86	49·50	53·59	55·83	57·24	58·20	58·91	59·44	59·86	60·19	60·71	61·22	61·74	62·00	62·26	62·53	62·79	63·06	63·33
2	8·53	9·00	9·16	9·24	9·29	9·33	9·35	9·37	9·38	9·39	9·41	9·42	9·44	9·45	9·46	9·47	9·47	9·48	9·49
3	5·54	5·46	5·39	5·34	5·31	5·28	5·27	5·25	5·24	5·23	5·22	5·20	5·18	5·18	5·17	5·16	5·15	5·14	5·13
4	4·54	4·32	4·19	4·11	4·05	4·01	3·98	3·95	3·94	3·92	3·90	3·87	3·84	3·83	3·82	3·80	3·79	3·78	3·76
5	4·06	3·78	3·62	3·52	3·45	3·40	3·37	3·34	3·32	3·30	3·27	3·24	3·21	3·19	3·17	3·16	3·14	3·12	3·10
6	3·78	3·46	3·29	3·18	3·11	3·05	3·01	2·98	2·96	2·94	2·90	2·87	2·84	2·82	2·80	2·78	2·76	2·74	2·72
7	3·59	3·26	3·07	2·96	2·88	2·83	2·78	2·75	2·72	2·70	2·67	2·63	2·59	2·58	2·56	2·54	2·51	2·49	2·47
8	3·46	3·11	2·92	2·81	2·73	2·67	2·62	2·59	2·56	2·54	2·50	2·46	2·42	2·40	2·38	2·36	2·34	2·32	2·29
9	3·36	3·01	2·81	2·69	2·61	2·55	2·51	2·47	2·44	2·42	2·38	2·34	2·30	2·28	2·25	2·23	2·21	2·18	2·16
10	3·29	2·92	2·73	2·61	2·52	2·46	2·41	2·38	2·35	2·32	2·28	2·24	2·20	2·18	2·16	2·13	2·11	2·08	2·06
11	3·23	2·86	2·66	2·54	2·45	2·39	2·34	2·30	2·27	2·25	2·21	2·17	2·12	2·10	2·08	2·05	2·03	2·00	1·97
12	3·18	2·81	2·61	2·48	2·39	2·33	2·28	2·24	2·21	2·19	2·15	2·10	2·06	2·04	2·01	1·99	1·96	1·93	1·90
13	3·14	2·76	2·56	2·43	2·35	2·28	2·23	2·20	2·16	2·14	2·10	2·05	2·01	1·98	1·96	1·93	1·90	1·88	1·85
14	3·10	2·73	2·52	2·39	2·31	2·24	2·19	2·15	2·12	2·10	2·05	2·01	1·96	1·94	1·91	1·89	1·86	1·83	1·80
15	3·07	2·70	2·49	2·36	2·27	2·21	2·16	2·12	2·09	2·06	2·02	1·97	1·92	1·90	1·87	1·85	1·82	1·79	1·76
16	3·05	2·67	2·46	2·33	2·24	2·18	2·13	2·09	2·06	2·03	1·99	1·94	1·89	1·87	1·84	1·81	1·78	1·75	1·72
17	3·03	2·64	2·44	2·31	2·22	2·15	2·10	2·06	2·03	2·00	1·96	1·91	1·86	1·84	1·81	1·78	1·75	1·72	1·69
18	3·01	2·62	2·42	2·29	2·20	2·13	2·08	2·04	2·00	1·98	1·93	1·89	1·84	1·81	1·78	1·75	1·72	1·69	1·66
19	2·99	2·61	2·40	2·27	2·18	2·11	2·06	2·02	1·98	1·96	1·91	1·86	1·81	1·79	1·76	1·73	1·70	1·67	1·63
20	2·97	2·59	2·38	2·25	2·16	2·09	2·04	2·00	1·96	1·94	1·89	1·84	1·79	1·77	1·74	1·71	1·68	1·64	1·61
21	2·96	2·57	2·36	2·23	2·14	2·08	2·02	1·98	1·95	1·92	1·87	1·83	1·78	1·75	1·72	1·69	1·66	1·62	1·59
22	2·95	2·56	2·35	2·22	2·13	2·06	2·01	1·97	1·93	1·90	1·86	1·81	1·76	1·73	1·70	1·67	1·64	1·60	1·57
23	2·94	2·55	2·34	2·21	2·11	2·05	1·99	1·95	1·92	1·89	1·84	1·80	1·74	1·72	1·69	1·66	1·62	1·59	1·55
24	2·93	2·54	2·33	2·19	2·10	2·04	1·98	1·94	1·91	1·88	1·83	1·78	1·73	1·70	1·67	1·64	1·61	1·57	1·53
25	2·92	2·53	2·32	2·18	2·09	2·02	1·97	1·93	1·89	1·87	1·82	1·77	1·72	1·69	1·66	1·63	1·59	1·56	1·52
26	2·91	2·52	2·31	2·17	2·08	2·01	1·96	1·92	1·88	1·86	1·81	1·76	1·71	1·68	1·65	1·61	1·58	1·54	1·50
27	2·90	2·51	2·30	2·17	2·07	2·00	1·95	1·91	1·87	1·85	1·80	1·75	1·70	1·67	1·64	1·60	1·57	1·53	1·49
28	2·89	2·50	2·29	2·16	2·06	2·00	1·94	1·90	1·87	1·84	1·79	1·74	1·69	1·66	1·63	1·59	1·56	1·52	1·48
29	2·89	2·50	2·28	2·15	2·06	1·99	1·93	1·89	1·86	1·83	1·78	1·73	1·68	1·65	1·62	1·58	1·55	1·51	1·47
30	2·88	2·49	2·28	2·14	2·05	1·98	1·93	1·88	1·85	1·82	1·77	1·72	1·67	1·64	1·61	1·57	1·54	1·50	1·46
40	2·84	2·44	2·23	2·09	2·00	1·93	1·87	1·83	1·79	1·76	1·71	1·66	1·61	1·57	1·54	1·51	1·47	1·42	1·38
60	2·79	2·39	2·18	2·04	1·95	1·87	1·82	1·77	1·74	1·71	1·66	1·60	1·54	1·51	1·48	1·44	1·40	1·35	1·29
120	2·75	2·35	2·13	1·99	1·90	1·82	1·77	1·72	1·68	1·65	1·60	1·55	1·48	1·45	1·41	1·37	1·32	1·26	1·19
∞	2·71	2·30	2·08	1·94	1·85	1·77	1·72	1·67	1·63	1·60	1·55	1·49	1·42	1·38	1·34	1·30	1·24	1·17	1·00

99th Percentiles

n_2 \ n_1	1	2	3	4	5	6	7	8	9	10	12	15	20	24	30	40	60	120	∞
1	161·4	199·5	215·7	224·6	230·2	234·0	236·8	238·9	240·5	241·9	243·9	245·9	248·0	249·1	250·1	251·1	252·2	253·3	254·3
2	18·51	19·00	19·16	19·25	19·30	19·33	19·35	19·37	19·38	19·40	19·41	19·43	19·45	19·45	19·46	19·47	19·48	19·49	19·50
3	10·13	9·55	9·28	9·12	9·01	8·94	8·89	8·85	8·81	8·79	8·74	8·70	8·66	8·64	8·62	8·59	8·57	8·55	8·53
4	7·71	6·94	6·59	6·39	6·26	6·16	6·09	6·04	6·00	5·96	5·91	5·86	5·80	5·77	5·75	5·72	5·69	5·66	5·63
5	6·61	5·79	5·41	5·19	5·05	4·95	4·88	4·82	4·77	4·74	4·68	4·62	4·56	4·53	4·50	4·46	4·43	4·40	4·36
6	5·99	5·14	4·76	4·53	4·39	4·28	4·21	4·15	4·10	4·06	4·00	3·94	3·87	3·84	3·81	3·77	3·74	3·70	3·67
7	5·59	4·74	4·35	4·12	3·97	3·87	3·79	3·73	3·68	3·64	3·57	3·51	3·44	3·41	3·38	3·34	3·30	3·27	3·23
8	5·32	4·46	4·07	3·84	3·69	3·58	3·50	3·44	3·39	3·35	3·28	3·22	3·15	3·12	3·08	3·04	3·01	2·97	2·93
9	5·12	4·26	3·86	3·63	3·48	3·37	3·29	3·23	3·18	3·14	3·07	3·01	2·94	2·90	2·86	2·83	2·79	2·75	2·71
10	4·96	4·10	3·71	3·48	3·33	3·22	3·14	3·07	3·02	2·98	2·91	2·85	2·77	2·74	2·70	2·66	2·62	2·58	2·54
11	4·84	3·98	3·59	3·36	3·20	3·09	3·01	2·95	2·90	2·85	2·79	2·72	2·65	2·61	2·57	2·53	2·49	2·45	2·40
12	4·75	3·89	3·49	3·26	3·11	3·00	2·91	2·85	2·80	2·75	2·69	2·62	2·54	2·51	2·47	2·43	2·38	2·34	2·30
13	4·67	3·81	3·41	3·18	3·03	2·92	2·83	2·77	2·71	2·67	2·60	2·53	2·46	2·42	2·38	2·34	2·30	2·25	2·21
14	4·60	3·74	3·34	3·11	2·96	2·85	2·76	2·70	2·65	2·60	2·53	2·46	2·39	2·35	2·31	2·27	2·22	2·18	2·13
15	4·54	3·68	3·29	3·06	2·90	2·79	2·71	2·64	2·59	2·54	2·48	2·40	2·33	2·29	2·25	2·20	2·16	2·11	2·07
16	4·49	3·63	3·24	3·01	2·85	2·74	2·66	2·59	2·54	2·49	2·42	2·35	2·28	2·24	2·19	2·15	2·11	2·06	2·01
17	4·45	3·59	3·20	2·96	2·81	2·70	2·61	2·55	2·49	2·45	2·38	2·31	2·23	2·19	2·15	2·10	2·06	2·01	1·96
18	4·41	3·55	3·16	2·93	2·77	2·66	2·58	2·51	2·46	2·41	2·34	2·27	2·19	2·15	2·11	2·06	2·02	1·97	1·92
19	4·38	3·52	3·13	2·90	2·74	2·63	2·54	2·48	2·42	2·38	2·31	2·23	2·16	2·11	2·07	2·03	1·98	1·93	1·88
20	4·35	3·49	3·10	2·87	2·71	2·60	2·51	2·45	2·39	2·35	2·28	2·20	2·12	2·08	2·04	1·99	1·95	1·90	1·84
21	4·32	3·47	3·07	2·84	2·68	2·57	2·49	2·42	2·37	2·32	2·25	2·18	2·10	2·05	2·01	1·96	1·92	1·87	1·81
22	4·30	3·44	3·05	2·82	2·66	2·55	2·46	2·40	2·34	2·30	2·23	2·15	2·07	2·03	1·98	1·94	1·89	1·84	1·78
23	4·28	3·42	3·03	2·80	2·64	2·53	2·44	2·37	2·32	2·27	2·20	2·13	2·05	2·01	1·96	1·91	1·86	1·81	1·76
24	4·26	3·40	3·01	2·78	2·62	2·51	2·42	2·36	2·30	2·25	2·18	2·11	2·03	1·98	1·94	1·89	1·84	1·79	1·73
25	4·24	3·39	2·99	2·76	2·60	2·49	2·40	2·34	2·28	2·24	2·16	2·09	2·01	1·96	1·92	1·87	1·82	1·77	1·71
26	4·23	3·37	2·98	2·74	2·59	2·47	2·39	2·32	2·27	2·22	2·15	2·07	1·99	1·95	1·90	1·85	1·80	1·75	1·69
27	4·21	3·35	2·96	2·73	2·57	2·46	2·37	2·31	2·25	2·20	2·13	2·06	1·97	1·93	1·88	1·84	1·79	1·73	1·67
28	4·20	3·34	2·95	2·71	2·56	2·45	2·36	2·29	2·24	2·19	2·12	2·04	1·96	1·91	1·87	1·82	1·77	1·71	1·65
29	4·18	3·33	2·93	2·70	2·55	2·43	2·35	2·28	2·22	2·18	2·10	2·03	1·94	1·90	1·85	1·81	1·75	1·70	1·64
30	4·17	3·32	2·92	2·69	2·53	2·42	2·33	2·27	2·21	2·16	2·09	2·01	1·93	1·89	1·84	1·79	1·74	1·68	1·62
40	4·08	3·23	2·84	2·61	2·45	2·34	2·25	2·18	2·12	2·08	2·00	1·92	1·84	1·79	1·74	1·69	1·64	1·58	1·51
60	4·00	3·15	2·76	2·53	2·37	2·25	2·17	2·10	2·04	1·99	1·92	1·84	1·75	1·70	1·65	1·59	1·53	1·47	1·39
120	3·92	3·07	2·68	2·45	2·29	2·17	2·09	2·02	1·96	1·91	1·83	1·75	1·66	1·61	1·55	1·50	1·43	1·35	1·25
∞	3·84	3·00	2·60	2·37	2·21	2·10	2·01	1·94	1·88	1·83	1·75	1·67	1·57	1·52	1·46	1·39	1·32	1·22	1·00

TABLE A2-3. **Percentage Points of the F-Distribution (*Continued*)** **99.5th Percentiles**

n_2 \ n_1	1	2	3	4	5	6	7	8	9	10	12	15	20	24	30	40	60	120	∞
1	16211	20000	21615	22500	23056	23437	23715	23925	24091	24224	24426	24630	24836	24940	25044	25148	25253	25359	25465
2	198·5	199·0	199·2	199·2	199·3	199·3	199·4	199·4	199·4	199·4	199·4	199·4	199·4	199·5	199·5	199·5	199·5	199·5	199·5
3	55·55	49·80	47·47	46·19	45·39	44·84	44·43	44·13	43·88	43·69	43·39	43·08	42·78	42·62	42·47	42·31	42·15	41·99	41·83
4	31·33	26·28	24·26	23·15	22·46	21·97	21·62	21·35	21·14	20·97	20·70	20·44	20·17	20·03	19·89	19·75	19·61	19·47	19·32
5	22·78	18·31	16·53	15·56	14·94	14·51	14·20	13·96	13·77	13·62	13·38	13·15	12·90	12·78	12·66	12·53	12·40	12·27	12·14
6	18·63	14·54	12·92	12·03	11·46	11·07	10·79	10·57	10·39	10·25	10·03	9·81	9·59	9·47	9·36	9·24	9·12	9·00	8·88
7	16·24	12·40	10·88	10·05	9·52	9·16	8·89	8·68	8·51	8·38	8·18	7·97	7·75	7·65	7·53	7·42	7·31	7·19	7·08
8	14·69	11·04	9·60	8·81	8·30	7·95	7·69	7·50	7·34	7·21	7·01	6·81	6·61	6·50	6·40	6·29	6·18	6·06	5·95
9	13·61	10·11	8·72	7·96	7·47	7·13	6·88	6·69	6·54	6·42	6·23	6·03	5·83	5·73	5·62	5·52	5·41	5·30	5·19
10	12·83	9·43	8·08	7·34	6·87	6·54	6·30	6·12	5·97	5·85	5·66	5·47	5·27	5·17	5·07	4·97	4·86	4·75	4·64
11	12·23	8·91	7·60	6·88	6·42	6·10	5·86	5·68	5·54	5·42	5·24	5·05	4·86	4·76	4·65	4·55	4·44	4·34	4·23
12	11·75	8·51	7·23	6·52	6·07	5·76	5·52	5·35	5·20	5·09	4·91	4·72	4·53	4·43	4·33	4·23	4·12	4·01	3·90
13	11·37	8·19	6·93	6·23	5·79	5·48	5·25	5·08	4·94	4·82	4·64	4·46	4·27	4·17	4·07	3·97	3·87	3·76	3·65
14	11·06	7·92	6·68	6·00	5·56	5·26	5·03	4·86	4·72	4·60	4·43	4·25	4·06	3·96	3·86	3·76	3·66	3·55	3·44
15	10·80	7·70	6·48	5·80	5·37	5·07	4·85	4·67	4·54	4·42	4·25	4·07	3·88	3·79	3·69	3·58	3·48	3·37	3·26
16	10·58	7·51	6·30	5·64	5·21	4·91	4·69	4·52	4·38	4·27	4·10	3·92	3·73	3·64	3·54	3·44	3·33	3·22	3·11
17	10·38	7·35	6·16	5·50	5·07	4·78	4·56	4·39	4·25	4·14	3·97	3·79	3·61	3·51	3·41	3·31	3·21	3·10	2·98
18	10·22	7·21	6·03	5·37	4·96	4·66	4·44	4·28	4·14	4·03	3·86	3·68	3·50	3·40	3·30	3·20	3·10	2·99	2·87
19	10·07	7·09	5·92	5·27	4·85	4·56	4·34	4·18	4·04	3·93	3·76	3·59	3·40	3·31	3·21	3·11	3·00	2·89	2·78
20	9·94	6·99	5·82	5·17	4·76	4·47	4·26	4·09	3·96	3·85	3·68	3·50	3·32	3·22	3·12	3·02	2·92	2·81	2·69
21	9·83	6·89	5·73	5·09	4·68	4·39	4·18	4·01	3·88	3·77	3·60	3·43	3·24	3·15	3·05	2·95	2·84	2·73	2·61
22	9·73	6·81	5·65	5·02	4·61	4·32	4·11	3·94	3·81	3·70	3·54	3·36	3·18	3·08	2·98	2·88	2·77	2·66	2·55
23	9·63	6·73	5·58	4·95	4·54	4·26	4·05	3·88	3·75	3·64	3·47	3·30	3·12	3·02	2·92	2·82	2·71	2·60	2·48
24	9·55	6·66	5·52	4·89	4·49	4·20	3·99	3·83	3·69	3·59	3·42	3·25	3·06	2·97	2·87	2·77	2·66	2·55	2·43
25	9·48	6·60	5·46	4·84	4·43	4·15	3·94	3·78	3·64	3·54	3·37	3·20	3·01	2·92	2·82	2·72	2·61	2·50	2·38
26	9·41	6·54	5·41	4·79	4·38	4·10	3·89	3·73	3·60	3·49	3·33	3·15	2·97	2·87	2·77	2·67	2·56	2·45	2·33
27	9·34	6·49	5·36	4·74	4·34	4·06	3·85	3·69	3·56	3·45	3·28	3·11	2·93	2·83	2·73	2·63	2·52	2·41	2·29
28	9·28	6·44	5·32	4·70	4·30	4·02	3·81	3·65	3·52	3·41	3·25	3·07	2·89	2·79	2·69	2·59	2·48	2·37	2·25
29	9·23	6·40	5·28	4·66	4·26	3·98	3·77	3·61	3·48	3·38	3·21	3·04	2·86	2·76	2·66	2·56	2·45	2·33	2·21
30	9·18	6·35	5·24	4·62	4·23	3·95	3·74	3·58	3·45	3·34	3·18	3·01	2·82	2·73	2·63	2·52	2·42	2·30	2·18
40	8·83	6·07	4·98	4·37	3·99	3·71	3·51	3·35	3·22	3·12	2·95	2·78	2·60	2·50	2·40	2·30	2·18	2·06	1·93
60	8·49	5·79	4·73	4·14	3·76	3·49	3·29	3·13	3·01	2·90	2·74	2·57	2·39	2·29	2·19	2·08	1·96	1·83	1·69
120	8·18	5·54	4·50	3·92	3·55	3·28	3·09	2·93	2·81	2·71	2·54	2·37	2·19	2·09	1·98	1·87	1·75	1·61	1·43
∞	7·88	5·30	4·28	3·72	3·35	3·09	2·90	2·74	2·62	2·52	2·36	2·19	2·00	1·90	1·79	1·67	1·53	1·36	1·00

Source: Reprinted from *Biometrika Tables for Statisticians* (2nd ed., 1962), eds. E. S. Pearson and H. O. Hartley, by permission of the *Biometrika* Trustees and the publisher, Cambridge University Press.

TABLE A2-4. **Percentiles of the Chi-Square Distribution**

d.f.	X^2_{1}	X^2_{2}	X^2_{5}	X^2_{10}	X^2_{20}	X^2_{30}	X^2_{50}	X^2_{70}	X^2_{80}	X^2_{90}	X^2_{95}	X^2_{98}	X^2_{99}	$X^2_{99.9}$
1	.00016	.00063	.0039	.016	.064	.15	.46	1.07	1.64	2.71	3.84	5.41	6.64	10.83
2	.02	.04	.10	.21	.45	.71	1.39	2.41	3.22	4.60	5.99	7.82	9.21	13.82
3	.12	.18	.35	.58	1.00	1.42	2.37	3.66	4.64	6.25	7.82	9.84	11.34	16.27
4	.30	.43	.71	1.06	1.65	2.20	3.36	4.88	5.99	7.78	9.49	11.67	13.28	18.46
5	.55	.75	1.14	1.61	2.34	3.00	4.35	6.06	7.29	9.24	11.07	13.39	15.09	20.52
6	.87	1.13	1.64	2.20	3.07	3.83	5.35	7.23	8.56	10.64	12.59	15.03	16.81	22.46
7	1.24	1.56	2.17	2.83	3.82	4.67	6.35	8.38	9.80	12.02	14.07	16.62	18.48	24.32
8	1.65	2.03	2.73	3.49	4.59	5.53	7.34	9.52	11.03	13.36	15.51	18.17	20.09	26.12
9	2.09	2.53	3.32	4.17	5.38	6.39	8.34	10.66	12.24	14.68	16.92	19.68	21.67	27.88
10	2.56	3.06	3.94	4.86	6.18	7.27	9.34	11.78	13.44	15.99	18.31	21.16	23.21	29.59
11	3.05	3.61	4.58	5.58	6.99	8.15	10.34	12.90	14.63	17.28	19.68	22.62	24.72	31.26
12	3.57	4.18	5.23	6.30	7.81	9.03	11.34	14.01	15.81	18.55	21.03	24.05	26.22	32.91
13	4.11	4.76	5.89	7.04	8.63	9.93	12.34	15.12	16.98	19.81	22.36	25.47	27.69	34.53
14	4.66	5.37	6.57	7.79	9.47	10.82	13.34	16.22	18.15	21.06	23.68	26.87	29.14	36.12
15	5.23	5.98	7.26	8.55	10.31	11.72	14.34	17.32	19.31	22.31	25.00	28.26	30.58	37.70
16	5.81	6.61	7.96	9.31	11.15	12.62	15.34	18.42	20.46	23.54	26.30	29.63	32.00	39.29
17	6.41	7.26	8.67	10.08	12.00	13.53	16.34	19.51	21.62	24.77	27.59	31.00	33.41	40.75
18	7.02	7.91	9.39	10.86	12.86	14.44	17.34	20.60	22.76	25.99	28.87	32.35	34.80	42.31
19	7.63	8.57	10.12	11.65	13.72	15.35	18.34	21.69	23.90	27.20	30.14	33.69	36.19	43.82
20	8.26	9.24	10.85	12.44	14.58	16.27	19.34	22.78	25.04	28.41	31.41	35.02	37.57	45.32
21	8.90	9.92	11.59	13.24	15.44	17.18	20.34	23.86	26.17	29.62	32.67	36.34	38.93	46.80
22	9.54	10.60	12.34	14.04	16.31	18.10	21.24	24.94	27.30	30.81	33.92	37.66	40.29	48.27
23	10.20	11.29	13.09	14.85	17.19	19.02	22.34	26.02	28.43	32.01	35.17	38.97	41.64	49.73
24	10.86	11.99	13.85	15.66	18.06	19.94	23.34	27.10	29.55	33.20	36.42	40.27	42.98	51.18
25	11.52	12.70	14.61	16.47	18.94	20.87	24.34	28.17	30.68	34.38	37.65	41.57	44.31	52.62
26	12.20	13.41	15.38	17.29	19.82	21.79	25.34	29.25	31.80	35.56	38.88	42.86	45.64	54.05
27	12.88	14.12	16.15	18.11	20.70	22.72	26.34	30.32	32.91	36.74	40.11	44.14	46.96	55.48
28	13.56	14.85	16.93	18.94	21.59	23.65	27.34	31.39	34.03	37.92	41.34	45.42	48.28	56.89
29	14.26	15.57	17.71	19.77	22.48	24.58	28.34	32.46	35.14	39.09	42.56	46.69	49.59	58.30
30	14.95	16.31	18.49	20.60	23.36	25.51	29.34	33.53	36.25	40.26	43.77	47.96	50.89	59.70

Source: Adapted from Table III of Fisher and Yates: *Statistical Tables for Biological, Agricultural and Medical Research,* Ed. 6. Hafner Publishing Co., 1963, with permission.

APPENDIX 3

Glossary of Symbols and Abbreviations

In the following glossary of symbols and abbreviations that are used throughout this book, Roman letters appear first, Greek letters next, and miscellaneous symbols at the end. Symbols that occur only in Chapter 11, or the meaning of a symbol that is specific to Chapter 11, are not included. Alternative symbols that are also often used, and which occur in some of the problems, are indicated in parentheses.

AR	attributable risk
b_0	sample intercept
$b_1, b_2, \ldots$	sample regression coefficients
d	deviation from the sample mean; difference between paired values
CV	coefficient of variation
d.f.	degrees of freedom
dl	deciliter
e	irrational mathematical constant equal to about 2.71828; expected number of counts in a table (also denoted E); estimated error or residual from a regression model
F	percentile of the F-distribution or the corresponding test statistic
f(y)	(probability) density function
F(y)	cumulative (probability) distribution function
g_2	fourth cumulant; the coefficient of kurtosis minus three
log	logarithm
$\log_e$	logarithm to base e; natural logarithm
Hg	mercury

H_0	null hypothesis
I	incidence
l	liter
mg	milligram
MS	mean square
n	sample size; parameter of the binomial distribution
o	observed number of counts in a table (also denoted O)
OR	odds ratio
p	sample proportion (also used to denote the probability parameter of the binomial distribution); p-value (also denoted P)
P(A)	probability of the event A
P(A|B)	conditional probability of the event A given B
r	correlation coefficient
R	multiple correlation coefficient
R^2	proportion of the variability explained by a regression model
RR	relative risk
s	sample standard deviation (estimate)
s^2	sample variance (estimate)
s_{xy}	sample covariance of X and Y (estimate)
S	sample standard deviation (estimator)
S^2	sample variance (estimator)
s.e.m.	standard error of the mean
SS	sum of squares
t	percentile of Student's t-distribution or the corresponding test statistic
T	rank sum statistic
x^2	sample chi-square statistic (also denoted X^2, χ^2)
x, y	particular values of (random) variables
X, Y	random variables
$\bar{y}$	sample mean of y (estimate)
$\bar{Y}$	sample mean of Y (estimator)
$\hat{y}$	value of y predicted from a regression equation
z	particular value of a standardized random variable
Z	standardized random variable
α	probability of type I error; significance level
β	probability of type II error; complement of power

β_0	population intercept
$\beta_1, \beta_2, \ldots$	population regression coefficients
ϵ	error or residual from a regression model
λ	parameter of the Poisson distribution
μ	population mean
π	irrational mathematical constant equal to the ratio of the circumference of a circle to its diameter; population proportion
σ	population standard deviation
σ^2	population variance
Σ	summation
χ^2	percentile of the chi-square distribution (also used to denote the corresponding statistic)
$=$	equals
$\doteq$	equals approximately
$\pm$	plus or minus
$<$	is less than
$\leq$	is less than or equal to
$>$	is greater than
$\geq$	is greater than or equal to
$/$	divided by, per
$\%$	percent
$\sqrt{\ }$	square root
$!$	factorial
∞	infinity

INDEX

A "t" following a page number indicates a table; an "f" indicates a figure.